Industrial Ecology

Industrial Ecology

Nataliya O. Smith

Delve
PUBLISHING

www.delvepublishing.com

Industrial Ecology

Nataliya O. Smith

Delve Publishing
2010 Winston Park Drive,
2nd Floor
Oakville, ON L6H 5R7
Canada
www.delvepublishing.com
Tel: 001-289-291-7705
001-905-616-2116
Fax: 001-289-291-7601
Email: orders@arclereducation.com

ISBN: 978-1-77361-001-6 (Hardcover)

Delve Publishing publishes wide variety of books and eBooks. For more information about Delve Publishing and its products, visit our website at www.delvepublishing.com.

ABOUT THE AUTHOR

Nataliya O. Smith

Dr. Nataliya O. Smith has extensive research, teaching and practical experience in Environmental Studies with accent on environmental safety, industrial ecology, and renewable energy sources as drivers of sustainable development. Over last ten years she has been working at Petro Mohyla Black Sea National University (Ukraine), where she had made path from PhD student and Adjunct Instructor to PhD, Associate Professor position in the Environmental and Natural Resources Department. Currently Mrs. Smith lives in the United States, where she is expanding her career.

Her PhD research, including numerous publications, has focused upon the study of environmental safety of heating-and-cooling-supply systems of buildings, the development of scientific methods of complex assessment of environmental impact and creating the mechanism for increasing environmental safety through sustainable use of renewable and traditional energy sources.

TABLE OF CONTENTS

LIST OF FIGURES

LIST OF TABLES

PREFACE

This book provides an overview of Industrial Ecology as a holistic study and a practical framework that serves to identify and then implement strategies to reduce the environmental impacts of manufacturing processes with an ultimate goal of inclusive and sustainable industrial development. Paper attempts to get readers ready to implement the natural ecosystem principles in industrial systems, and to initiate discussion on the direction along which the Industrial Ecology approach could be developed.

Studying the background of Industrial Ecology, its main objectives and key concept, this work is trying to analyze the main types of industrial environmental impacts, their causes and consequences, investigate how different industrial sectors influence the environmental and human health, and discover what new technologies we have to eliminate that hazardous effects. And eventually, the last chapter uncovers the most severe industrial environmental disasters of the past, so we will always remember them and never let such tragedies danger the humanity again.

The book could be useful for students who study Environmental Science and Engineering, Environmental Management, Urban Studies, industrial engineers, environmental policy planners, urban designers, local authorities, public organizations, and active communities.

Author

1 CHAPTER

OVERVIEW OF INDUSTRY

INDUSTRY DEFINITION

There are some different approaches to define what is industry around the world. According to the Business dictionary, "Industry is the manufacturing or technically productive enterprises in a particular field, country, region, or economy viewed collectively, or one of these individually. A single industry is often named after its principal product; for example, the auto industry"; or "Industry is any general business activity or commercial enterprise that can be isolated from others, such as the tourist industry or the entertainment industry" (Business Dictionary, 2013). Cambridge dictionary proposes such definitions: "industry is the aggregate of manufacturing or technically productive enterprises in a particular field, often named after its principal product" and "the aggregate of work, scholarship, and ancillary activity in a particular field, often named after its principal subject", stressing attention on manufacturing and technological aspects of industrial process (Cambridge Dictionary, 2010).

United Nations refers industry to the "activity of the establishment in which the person worked or was employed during the reference period"

(UN, 2012). The European System of Accounts (ESA) defines the term 'Industry' as "consisting of a group of local kinds of activity units engaged in the same, or similar, kind-of-activity. At the most detailed level of classification, an industry consists of all the local kinds of activity units falling within a single class and which are therefore engaged in the same type of productive activity, irrespective of whether or not the institutional units to which they belong produce market or non-market output" (OECD, 2013).

Post-Soviet Eastern European countries understand industry as "technically the most perfect field of material production, the basis of industrialization of the economy, which has a decisive influence on the development of productive forces; a set of enterprises producing electricity, tools for economic sectors, extraction of raw materials, fossil fuels, logging, processing of agricultural products, production of goods and services" (Shevchenko, 2008).

As we see, there are two different points of view on the industry – economical and manufacturing. Business statistics still doesn't have harmonized definition for the term 'Industry' due to its wide usage in different conditions. Economists usually use 'industry' as a synonym to 'activity', to 'industrial sector' and to 'industrial activity'. Manufacturers, on the other hand, refers industry to 'production' and 'technology'. But living in a fast-changing world, today we should understand industry in context of sustainable development that pursues the socially inclusive industrial development, that offer equal opportunities and an equitable distribution of benefits, taking in account environmental sustainability, decoupling the prosperity generated by industrial activities from excessive natural resource use and minimal or even zero environmental impact.

INDUSTRY CLASSIFICATION

In modern economies, there are dozens of industry classifications, which are typically grouped into larger categories called sectors, with individual companies being classified into an industry based on their largest sources of revenue (Investopedia, 2009). According to the Sector theory, industry is usually classified into sectors:

- Primary – Extractive industry;
- Secondary – Manufacturing industry;
- Tertiary – Services.
- Quaternary – Knowledge;
- Quinary – Culture and research.

During historical development, the segmentation of a society's industry within each sector can change. At the Figure 1 we can analyze the GDP composition of sector and labor force by occupation in the form of any component to economy. The green, red, and blue components of the colors of the countries represent the percentages for the agriculture, industry, and services sectors, respectively. While analyzing, please, note: map colors are more finely graded than key colors (Wikipedia, 2017).

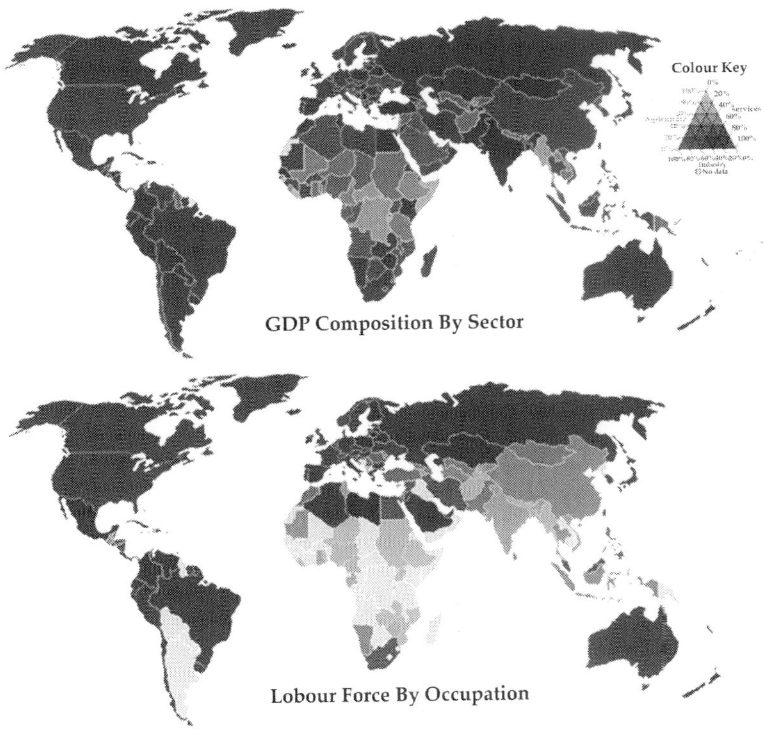

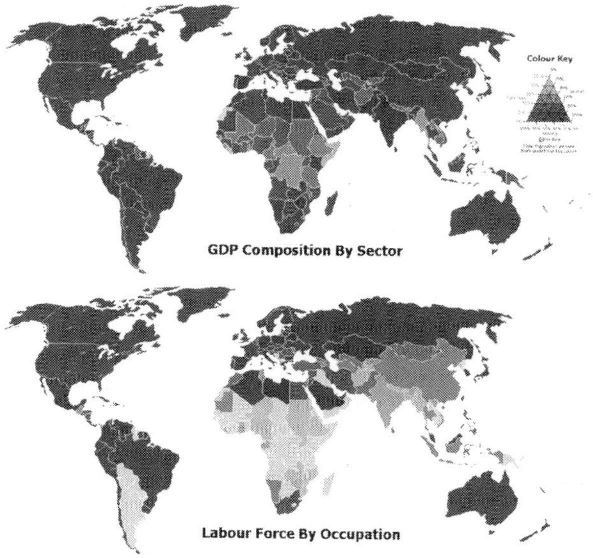

Figure 1: GDP Composition by Sector and Labor Force by Occupation in the World (2015). Source: Wikipedia, (2017). Available at: https://en.wikipedia.org/wiki/Industry#/media/File:Gdp-and-labour-force-by-sector.png

Industry can also be classified by main product of manufacturing sectors, such as:

- Construction/building industry;
- Chemical industry;
- Fuel/petroleum industry;
- Automotive industry;
- Electronic industry;
- Power engineering/manufacturing industry;
- Meatpacking industry;
- Hospitality industry;
- Food industry;
- Fish industry;
- Software/IT industry;
- Paper industry;
- Entertainment industry;
- Semiconductor industry;

- Cultural industry;
- Poverty industry.

Another industrial classification system takes in account main technological processes used in industrial production (Figure 2)

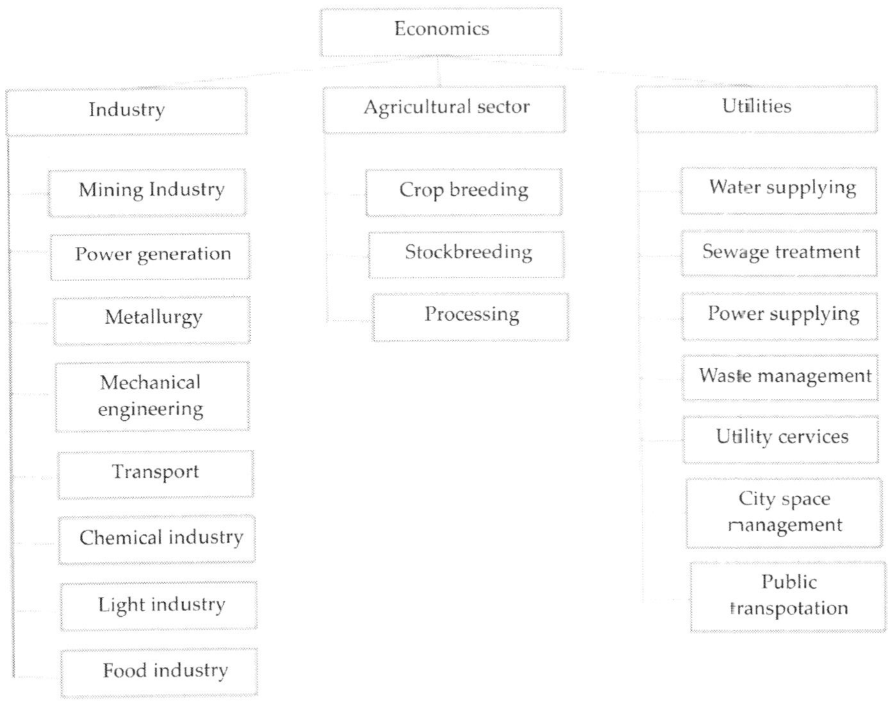

Figure 2: Economics Classification Considering the Industrial Processes.

Main economic/industrial sectors can be classified in more detailed way. These classification systems commonly divide industries according to similar functions and markets and identify businesses producing related products. Market-based classification systems such as the Global Industry Classification Standard and the Industry Classification Benchmark are used in finance and market explorations. The International Standard Industrial Classification (ISIC) of all economic activities is considered to be the most complete and systematic industrial classification made by the United Nations Statistics Division (OECD, 2013). ISIC is a standard classification of economic activities arranged so that entities can be classified according to the activity they carry out. The categories of ISIC at the most detailed level (classes) are delineated according to what is, in most countries, the customary combination of activities described in statistical units, considering the relative importance of the activities included in these classes (OECD, 2013).

The North American Industry Classification System (NAICS), developed and broadly used by the United States, Canada and Mexico, is widely used by investors to classify companies and is the standard upon which government agencies classify businesses when compiling statistical data. In the NAICS hierarchy, companies that use similar manufacturing processes are characterized in the same industry (Investopedia, 2009).

INDUSTRIAL DEVELOPMENT

Technology is a driving force of industrial development. The history of nascence, accumulation and changes in technology is interesting and useful not only because it traces the growth of human power, but also because it is the only source that directly connects the ancient times to the present. Analyzing the human development, we can distinguish individual links of the technology chain history. Joining each new level enriches the arsenal of humanity in our relationship with nature. Figuratively speaking, people gradually enfetter nature with technological chain. Among the technological chain links, there are some, which have had a revolutionary impact not only on the development of material culture, but also on human society in general. These technological steps (or even, revolutions) during our development can be formed in the Technology chain of humanity:

- **First** – Creation of artificial means of production (sharped sticks, stones).
- **Second** – Mastering the artificial heat (friction technology) and using it in the industrial processes.
- **Third** – Transition from hunting and plant gathering to agriculture and breeding.
- **Fourth** – Exploitation of animal energy in technological processes (elephants, oxen, horses).
- **Fifth** – Exploitation of inanimate natural energy in industrial processes (water, wind).

- **Sixth** – Beginning of usage of technological equipment (replacement of hand function).
- **Seventh** – Mastering the technology of converting the thermal energy into the mechanical energy (Industrial revolution – invention of steam engine).
- **Eighth** – Mastering the technology of material production with chemical transformation.
- **Ninth** – Mastering the nuclear energy.
- **Tenth** – Mastering the technology of material production with physical transformation (synthetic materials, Nano-technology).
- **Eleventh** – Development of Information Technology (automatic control systems, INTERNET, mobile technology).
- **Twelfth** – Mastering the biological technology (genetic engineering).
- **Thirteenth** – Mastering the space technology, usage of space resources (SpaceX, Mars Mission).
- **Fourteenth** – Creation of artificial intelligence.

Each of these process units can be accompanied by the large number of examples to prove their epoch-making significance. Technology chain of humanity doesn't have the end, it will attach more and more links while mankind exists.

Officially industrial development began with Industrial Revolution and invention of steam engine, which allowed to increase the intensity of industrial processes dramatically. Manufacturing industry became a key sector of production and labor in European and North American countries, upsetting previous mercantile and feudal economies. This came through many successive rapid advances in technology, such as the production of steel and coal.

A factory is a traditional symbol of the industrial development. The Industrial Revolution led to the development of factories for large-scale production, with consequent changes in society and in environmental impact. Originally the factories were steam-powered, but later transitioned to electricity once an electrical grid was developed. The mechanized assembly line was introduced to assemble parts in a repeatable fashion, with individual workers performing specific steps during the process. This led to significant increases in efficiency, lowering the cost of the end process. Later automation was increasingly used to replace human operators. This process has accelerated with the development of the computer and the robot. And every that step was accompanied by intensifying of resource- usage with hazardous effects on natural environment and human health (Wikipedia, 2017).

After the Industrial Revolution, conceivably a third of the world's economic output are derived from manufacturing industries. Many developed countries and many developing/semi-developed countries, such as China, India etc., depend fundamentally on manufacturing industry. Industries, the countries they reside in, and the economies of those countries are interconnected in a complex net of interdependence (Wiedmann et.al., 2013).

During the time, due to various economic factors, including the development of replacement technology or the loss of competitive advantage, certain manufacturing industries have gone into a decline. An example of the former is the decline in cable phones and telegraphs manufacturing when the mobile mass-connection was introduced.

The migration of prosperous has been a recent trend in modern world; industrialized nations move in the direction of a postindustrial society. This is revealed by a growth of the service sector at the expense of manufacturing, and the development of an information-based economy, the so-called informational revolution. In a postindustrial society, manufacturing is relocated to economically more favorable locations through a process of off-shoring (DocSlides, 2016). Figure 3 shows us the distribution of countries in the world by industrial output in 2015 according to IMF and CIA World Factbook.

Countries by industrial output in 2015 (billions in USD)

Economy	
(01) China	4,922
(_) European Union	4'162
(02) United States	3,752
(03) Japan	1,062
[illegible]	1,051
(05) United Kingdom	588
(06) India	559
(07) South Korea	555
(08) Brazil	496
(09) Canada	456
(10) France	453
(11) Mexico	448
(12) Italy	440
(13) Russia	427
(14) Indonesia	408
(15) Saudi Arabia	387
(16) Australia	362
(17) Spain	312
(18) United Arab Emirates	214
(19) Turkey	202
(20) Switzerland	184

Figure 3: Distribution of Countries in the World by Industrial Output in 2015.

Source: The World Factbook https://www.cia.gov/library/publications/resources/the-world-factbook/

Latest PwC report have concluded that about a third of jobs in the US, Germany and Britain could be eliminated by automation by the early 2030s, with the losses concentrated in transportation and storage, manufacturing, and wholesale and retail trade (PwC, 2017). Their other report sets out long-term global growth projections to 2050 for 32 of the largest economies in the world, accounting for around 85% of world GDP:

- The world economy could more than double in size by 2050, far outstripping population growth, due to continued technology-driven productivity improvements;
- Emerging markets (E7) could grow around twice as fast as advanced economies (G7) on average (Figure 4);
- As a result, six of the seven largest economies in the world are expected to be emerging economies in 2050 leading by China (1st), India (2nd) and Indonesia (4th);
- The United States could be down to third place in the global GDP rankings while the EU27's share of world GDP could fall below 10% by 2050;

- UK could be down to 10th place by 2050, France could lose its position in the top 10 and Italy could lose its position in the top 20 as they are overtaken by faster growing emerging economies like Mexico, Turkey and Vietnam respectively;
- Emerging economies need to enhance their institutions and their infrastructure significantly if they are to realize their long-term growth potential (PwC 2017).

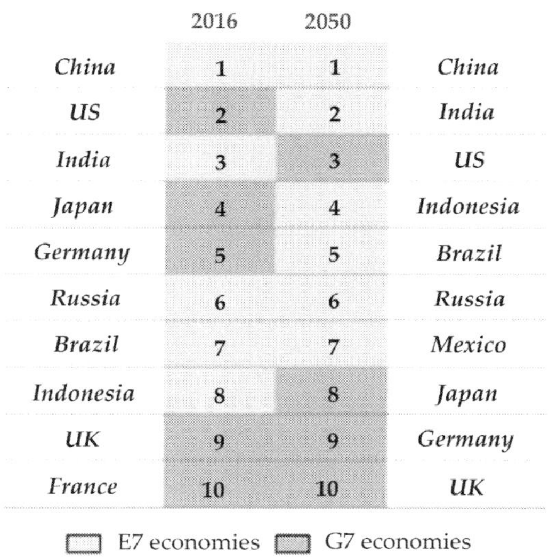

Figure 4: Projected Domination of Emerging Markets in the World's Top Ten Economies in 2050.

Source: PwC analysis for projections to 2050 https://www.pwc.com/gx/en/issues/economy/the-world-in-2050.html

As Bob Moritz, the PwC Chairman, mentioned, "It isn't often that the broad infrastructure that underlies industrial civilization undergoes a dramatic transformation. But just such a change appears to be happening now. As sensors spread through factories and utility services, power grids and loading docks become intelligent, and custom-designed products are created on demand, the label the next industrial revolution is coming to represent this great wave of technological change. The leaders of this revolution are companies making advances in fields such as robotics, machine learning, digital fabrication (including 3D printing), the Internet of Things (IoT), data analytics and blockchain (a system of decentralized, automated transaction verification). Because these technologies all reinforce the others' impact, they are leading to a new level of proficiency, and new types of opportunities and challenges for industry and for society at large" (PwC, 2017).

Along with such dramatic changes in industrial development, traditional boundaries between industries are eroding. It's getting harder to tell the difference between, for example, a telecommunications company and an entertainment producer, or between a retail bank and a retail store. The relationships among suppliers, producers, and consumers are obscuring very rapidly (Schwieters and Moritz, 2017).

Today United Nations (2017) proposes new model of industrial development – Inclusive Sustainable Industrial Development (ISID) that has three elements as a framework. The first is long-term, sustained industrialization as a driver for economic development. The second is socially inclusive industrial development that offers equal opportunities and an equitable distribution of benefits for every member of society. The third element is environmental sustainability, decoupling the prosperity generated by industrial activities from excessive natural resource use and negative environmental impact. This three-dimensional system feeds through to the policy recommendations for dealing with the many trade-offs that countries face in sustaining economic growth, promoting social inclusiveness, and moving towards greener economic transformations (UNIDO, 2017).

Manufacturing will play a key role in the long-term structural industrial change: it can create many productive, formal jobs at the early stage of development and also can drive technological innovation to sustain productivity growth. Manufacturing process has varying effects on employment, wages, technological upgrading, and sustainability at different stages of development. That happens because manufacturing tends to transform economic structures, usually from labor-intensive activities to more capital- and technology intensive activities. Each manufacturing subsector also changes products and production processes, with the increasing applications of capital and technology (UNIDO, 2017).

Even though technology has long affected the labor force, recent innovations in artificial intelligence and robotics can speed up the automation replacing of occupations, including highly skilled or 'knowledge-based' jobs. For example, algorithms play an increasing role in journalism today, medicine adapts robotic surgery and artificial intelligence to detect cancer and heart conditions, self-driving and drone technology may eliminate the need for taxi, Uber and lorry drivers (The Express Tribune, 2017).

Historian Yuval Harari writes in his 2016 book "Homo Deus: A Brief History of Tomorrow" that technology will lead to 'superfluous people' as 'intelligent non-conscious algorithms' improve. "As algorithms push humans out of the job market," he writes, "wealth and power might become concentrated in the hands of the tiny elite that owns the all-powerful algorithms, creating unprecedented social and political inequality" (Harari, 2016).

But not everybody agrees with such a dramatic prospective. Boston University economist and researcher James Bessen dismisses alarmist predictions, contending that advances in technology generally lead to more jobs, even if the nature of work changes. His research found that the proliferation of ATM machines did not decrease bank tellers' employment in recent decades, and that automation of textile mills in the 19th century led to an increase in weaving jobs because it created more demand. "Robots can replace humans in certain tasks but don't entirely replace humans," he says. But he acknowledged that automation "is destroying a lot of low-skill, low wage jobs and the new jobs being created need higher skills" (TheNational Business, 2017).

UN economists implied that premature deindustrialization could be a serious threat to sustainable growth in developing countries, smothering the growth potential of manufacturing when it sets in. These informal service activities, which emerge at this stage, slow the growth down rather than enhance it. But when deindustrialization sets in at higher levels of per capita income, the kind of services that emerge—coordination, business services and information technology services – are much more dynamic and can take over and complement the growth-enhancing role of manufacturing (UNIDO, 2017).

One statement has no doubt: technological change is recognized as one of the main drivers of long-term growth. The Sustainable Development Goal 9 "Build resilient infrastructure, promote inclusive and sustainable industrialization and foster innovation", adopted on 26 September

2015 Paris Agreement, argues that without technology and innovation, industrialization will not happen, and without industrialization, development will not happen. It is undebatable that technology makes manufacturing processes more efficient, thereby increasing the competitiveness of countries and reducing their vulnerability to market fluctuations. Structural change in industrial growth, i.e. the transition from a labor-intensive to a technology-intensive economy, drives economic advancement (UN, 2017). Our task today is to form inclusive and sustainable industrial development, which ensure wealth being and environmental safety for future generations. That is why implementation of Industrial Ecology principles is very important in planning and realization of the structural technological change and the long-term growth.

2
CHAPTER

INTRODUCTION TO INDUSTRIAL ECOLOGY

DEFINING THE INDUSTRIAL ECOLOGY

As we have learnt in the previous chapter, the global industry can be modelled as a network of industrial processes that extract resources from the Earth and transform those resources into commodities, which can be bought and sold to meet the needs of humanity. And one of the most important factors forming the inclusive and sustainable industrial development, is the implementation of Industrial Ecology principles and methods.

Before we come to definition what the Industrial Ecology is, we should analyze and understand where we need to place Humanity in its relations with Nature. Sustainable Development of Humanity should be based on Neo-Anthropocentric approach, which is founded on the research and analysis of the dualism of human being shown in Figure 5 (Dobrovolsky, 2012).

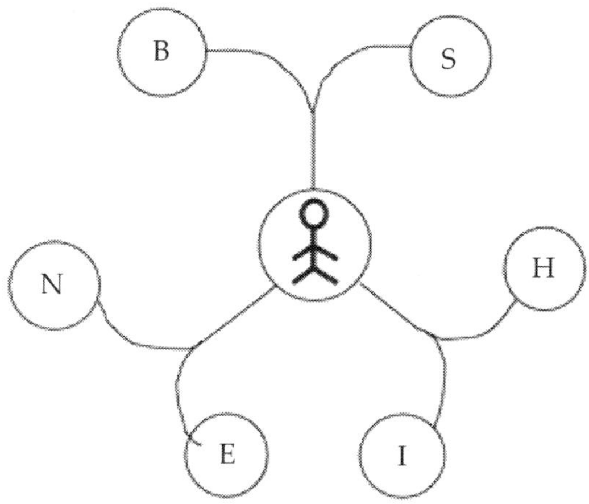

Figure 5: Dualism of Human Being: Body (B) – Soul (S); Individual (I) – Society / Humanity (H); Natural Creature (N) – Economical Actor / Resource User (E).

Source: Dobrovolsky, (2012).

Referring to the fundamental vision of the United Nations, announced in the Principle 1 "Human beings are at the center of concerns for sustainable development" (UN, 1992), we should mention differences in understanding the term 'anthropocentrism', which means the central position of humanity in the biosphere. During the industrialization, anthropocentrism was seen as the people right to meet their needs through overconsumption, ignoring the interests of other living creatures and abiotic components of biosphere.

The author of Biosphere Concept, Volodymyr Vernadsky, assigned to humanity the central place in the biosphere not from the standpoint of the rights, but from the position of responsibility for the fate of the biosphere – humanity with using the intelligence must evolve the biosphere into the noosphere. (Vernadsky 1926) In the UN documents, the terms 'biosphere', 'noosphere', 'anthropocentrism' are not used, but according to the content, they are talking about responsibility, not about special rights for the people. On the contrary, the main thought is about the limitation of human needs, the necessity in taking care of our common future and so on. Movement towards the sustainable development is impossible without a significant change in human behavior in the environment – both natural environment and social environment. From this point of view, it seems logical to assume 'the transformation of human biospheric being to human noospheric being' as the goal of sustainable development. Such neo-anthropocentrism of postindustrial society lays primarily in the self-improvement and self-cultivation of humanity. And Industrial Ecology is one of the instruments which should be used in the process of self-improvement of humanity, bringing new concepts and structural changes to industrial development.

Industrial Ecology is a young but growing multidisciplinary field of research which combines aspects of engineering, economics, sociology, toxicology, and the natural sciences. Different scientists propose different definitions of Industrial Ecology. In an emerging field of theory and practice it is natural that understanding of its scope and meaning should be varied. The Industrial Ecology field was officially created at the National Academy of Engineering meeting in 1992 with support from the AT&T Foundation. The International Society for Industrial Ecology was founded in 2001 (ISIE, 2015).

The term 'Industrial Ecology' was originated from the idea that the analogy of natural systems should be used as an assistance in understanding how to design sustainable industrial systems. (Frosch and Gallopoulos, 1989). Industrial Ecology is 'industrial' as it views manufacturers as agents for environmental improvement because they possess the technological expertise that is critical to the successful execution of sustainable design of products and processes (Graedel, 1996).

Industrial Ecology is 'ecological' based on two reasons. First, Industrial Ecology looks at non-artificial 'natural' ecosystems as potential models for industrial activity ('industrial ecosystem'). This is what some environmentalists called the 'biological analogy'. Natural ecosystems have the highest existing level of efficiency in energy and resource cycles; and many of them are especially effective at recycling resources and therefore are held out as ideal exemplars for efficient cycling of materials and energy in industry and society. Industrial Ecology also derives biological concepts related to the idea of metabolism. Second, Industrial Ecology places industry, as a widest sense of human technological activity, in the framework of the larger ecosystems which support it, examining the sources of resources used by society and the sinks that may act to absorb or recycle wastes. This concluding sense of 'ecological' links Industrial Ecology to questions of carrying capacity and environmental safety. That is why, this study has a strong anchor in engineering, but includes substantial elements of environmental science, social ecology, environmental economics, and management (Graedel, 1996).

Robert White, president of the U.S. National Academy of Engineering (White, 1994) defined: Industrial Ecology as "a study of the flows of materials and energy in industrial and consumer activities, of the effects of these flows on the environment, and of the influences of economic, political, regulatory, and social factors on the flow, use, and transformation of resources".

Andy Garner and Gregory Keoleian proposed such definition: "Industrial Ecology is the study of the physical, chemical, and biological interactions and interrelationships both within and between industrial and ecological systems" (Garner and Keoleian, 1995). Additionally, some researchers feel that "Industrial Ecology involves identifying and implementing strategies for industrial systems to more closely emulate harmonious, sustainable, ecological ecosystems" (Frosch, 1992).

According to Brattebo, "Industrial ecology can be viewed as a set of notions (theoretical/conceptual) and methods (practical/instrumental), which are using the natural eco-systems as a metaphor and model to express how industrial society could be organized and function" (Brattebo et. al., 1998). According to Lifseth (1997), industrial ecology can be seen as "the operational part of sustainable development. Brad Allenby looked at the Industrial Ecology as on a "systems-based, multidisciplinary discourse that seeks to understand emergent behavior of complex integrated human-natural systems" (Allenby, 2006).

FIELD OF STUDY

Industrial Ecology focuses on the interrelationships among manufacturers, as well as among their products and processes, at the local, regional, national, and global system levels. These layers of overlapping connections are similar to the food web that characterizes the interrelatedness of organisms in natural ecosystems. Industrial ecology has the close relationship with Applied Ecology and Social Ecology (Garner and Keoleian, 1995).

Analyzing how the *Journal of Applied Ecology* defines 'Applied Ecology', – "application of ecological ideas, theories and methods to the use of biological resources in the widest sense. It is concerned with the ecological principles underlying the management, control, and development of biological resources for agriculture, forestry, aquaculture, nature conservation, wildlife and game management, leisure activities, and the ecological effects of biotechnology", – we can see, that Industrial Ecology can use all the methods of Applied Ecology trying to transmit the principles and processes of natural ecosystems to develop models for efficient industrial systems (Garner and Keoleian, 1995).

The Institute of Social Ecology's definition of Social Ecology states that: "Social Ecology integrates the study of human and natural ecosystems through understanding the interrelationships of culture and nature. It advances a critical, holistic world view and suggests that creative human enterprise can construct an alternative future, reharmonizing people's relationship to the natural world by reharmonizing their relationship with each other" (Begens, et.al., 1991). Industrial Ecology must use this approach achieving the goal of building the new structure of inclusive sustainable industrial growth.

Thus, Industrial Ecology is a dynamic holistic framework that ensures the management of human manufacturing activity on a sustainable basis by:

- Minimizing the environmental effects of human activity to the level that natural systems can sustain;
- Ensuring environmental safety and acceptable quality of life for people;
- Improving energy and materials usage efficiency;
- Conserving and restoring ecosystem health and preserving biodiversity;
- Maintaining the economic viability of systems for industry, trade, and commerce (Indigo Development, 2005).

The U.S. Department of Labor's O*NET (Occupational Information Network) program describes the 'New and Emerging' occupation of

Industrial Ecologists and their work (NCOD, 2015): "Apply principles and processes of natural ecosystems to develop models for efficient industrial systems. Use knowledge from the physical and social sciences to maximize effective use of natural resources in the production and use of goods and services. Examine societal issues and their relationship with both technical systems and the environment."

According to NCOD (2015), tasks undertaken by industrial ecologists include:

- Analyzing the changes and designing the measures to improve the environmental performance of complex systems to avoid unintended negative consequences;
- Applying the new or existing research about natural ecosystems to understand economic and industrial systems in the context of the environment;
- Carrying out the environmental assessments in accordance with appropriate standards, regulations, or laws;
- Conducting analyses to determine the maximum amount of work that can be accomplished for a given amount of energy in a system, such as industrial production systems and waste treatment systems;
- Developing the alternative energy investment scenarios to compare economic and environmental costs and benefits;
- Evaluating the effectiveness of industrial ecology programs using statistical analysis and applications;
- Examining the local, regional, or global use and flow of materials or energy in industrial production processes;
- Examining the societal issues and their relationship with both technical systems and the environment.
- Identifying and equating the component parts or relationships between the parts of industrial, social, and natural systems;
- Identifying and developing the strategies or methods to minimize the environmental impact of industrial production processes (NCOD 2015).

Industrial Ecology tries to understand the way in which industrial systems (for example a factory, an ecoregion, or national or global economy) interact with the biosphere. Natural ecosystems provide a metaphor for understanding how different parts of industrial systems interact with one another, in an 'ecosystem' based on resources and infrastructural capital rather than on natural capital. The study seeks to exploit the idea that natural ecosystems do not produce and collect waste in them to inspire sustainable design (Wikipedia, 2017).

As a practical field, Industrial Ecology is concerned with the shifting of industrial process from linear (open loop) systems, in which resource and capital investments move through the system to become waste, to a closed loop system where wastes can become inputs for new processes (Butnariu and Avasilcai, 2013).

The Industrial Ecology approach involves:

- application of systems science to industrial systems;
- defining the system boundary to incorporate the natural world;
- seeking to optimize that system.

In this context, 'Industrial systems' applies not just to private sector manufacturing and service but also to government operations, including provision of infrastructure.

The main research areas of Industrial Ecology are:

- Material and energy flow analysis ('industrial metabolism');
- Dematerialization and decarbonization;
- Technological change and the environmental impact;
- Life-cycle planning, design, and assessment;
- Environmental design ('eco-design');
- Extended producer responsibility ('product stewardship');
- Eco-industrial parks ('industrial symbiosis');
- Product-oriented environmental policy;
- Eco-efficiency (Indigo Development, 2005; Butnariu and Avasilcai, 2013).

Along with more general energy efficiency and resource management goals, and redefining commodity markets and product stewardship relations strictly as a service economy, Industrial Ecology is one of the four objectives of Natural Capitalism. This strategy discourages forms of amoral purchasing arising from ignorance of what goes on at a distance and implies a political economy that values natural capital highly and relies on more instructional capital to design and maintain each unique Industrial Ecology (IIBM, 2015).

GOALS OF INDUSTRIAL ECOLOGY

The primary goal of Industrial Ecology is to promote sustainable development at the global, regional, and local levels. Sustainable development is defined by the United Nations World Commission on Environment and Development as "meeting the needs of the present generation without sacrificing the needs of future generations" (UN, 1992).

The challenge, according to Brattebo and his colleagues, is to develop methods, models and tools based on ecological principles and in harmony with nature's carrying capacity for implementation of preferred change. A new perspective, compared with other fields working on environmental problems, and the main approach in industrial ecology, is to combine economic- and environmental efficiency on both manufacturer- and society level (Brattebø et.al., 1998).

Key objectives of Industrial Ecology inherent to sustainable development include (Garner, Keoleian, 1995):

- Sustainable use of resources and energy;
- Preserving ecological (environmental) and human health (support of ecosystem structure and functions);
- Promotion of environmental equity (both intergenerational and inter-societal).

Let us describe these main objectives more detailly.

Sustainable Use of Resources: Industrial Ecology encourages the sustainable use of renewable energy and resources, and endeavor to minimal use of nonrenewable ones. Industrial activity is dependent on a stable supply of resources and thus should function with the highest possible level of efficiency. In the past humanity has found alternatives to diminished raw materials, it cannot be assumed that alternatives will continue to be found as supplies of primary raw materials decrease or are degraded. We must understand that supply of all natural resources is finite. Thus, depletion of non-renewables and degradation of renewables must be minimized for industrial activity sustainability in the long term (Garner and Keoleian, 1995).

Environmental and Human Health: Human beings are only the one component in a complex and diverse web of ecosystem interactions: their activities cannot be separated from the functioning of the entire system. Human health is dependent on the health of other ecosystem components and the level of environmental safety, that is why healthy ecosystem structure and function should be in the focus of Industrial Ecology. It is important that industrial activities should not cause catastrophic disruptions to the natural ecosystems or slowly degrade their structure and functions, jeopardizing the planet's life support system (Garner and Keoleian, 1995).

Environmental Equity: One of the primary challenges of sustainable development is the achieving of intergenerational as well as inter-societal equity. Exhausting natural resources and degrading environmental health to meet short-term economic objectives can threaten the environmental safety of future generations and their ability to meet their needs. Inter-societal discriminations also exist between developing and developed countries, as evidenced by the large disparity of resource use. Developed countries usually use a disproportionate bigger amount of resources in comparison with developing countries. We can also observe injustices between social and economic groups within one country. Several studies have shown that low income and ethnic communities in the U.S., for example, are often subject to much higher levels of human health risk related to certain toxic pollutants (Wiedmann and others 2013).

That is why one of the major objectives of Industrial Ecology is to influence industrial decision making grounded on sustainability concepts based on closing production loops in product life cycles, networks and food-webs, bio-mimicry that informs the design of eco-industrial parks, design-for-the-environment procedures for use in industrial conditions, material balancing for life-cycle assessments of products (Chertow, 1998; Graedel and Allenby, 1995; Winter, 1988; Indigo Development, 2003).

To sum up, the challenge of Industrial Ecology is to turn the individual business from the today's short-term micro-oriented viewpoint to a long-term global-oriented perspective, where environmental challenges are evaluated at the same level as the financial and social ones.

3
CHAPTER

KEY CONCEPTS OF INDUSTRIAL ECOLOGY

ECOSYSTEM PRINCIPLES IN INDUSTRIAL ECOSYSTEMS

As we have analyzed in previous chapter, Industrial Ecology approach is based on scientific efforts to mimic the conditions and processes of the natural ecosystem in an industrial system. In natural ecosystems, waste equals food, and the energy is flowed along the food chain while the only input to the system is the solar energy from the sun. The basic idea in the industrial ecosystem is to enhance the development of an industrial system that relies on co-operation between the actors involved, in that they use each other's waste material and energy as resources. Such approach allows to minimize the system virgin material and energy input, as well as the waste and emission output. It is obvious that a perfect industrial ecosystem will never happen. But it also seems to be clear that the direction we should follow when striving toward sustainable development is to learn from the ecosystem, even though this effort may still be only at a metaphoric level (Korhonen, 2001).

Industrial ecology recommends to use the strategy of ecosystems to guide the redesign of industrial systems. The goal is to achieve a better match of industrial performance with ecological limitations. The widest application of this analogic approach is to describe manufacturing complexes as 'industrial ecosystems' (Lowe and Evans, 1995). This idea suggests a web of interaction among companies such that the wastes and emissions of one facility become feedstock for another. Design of environmentally friendly production strategies then moves beyond the individual plant or company into the realm of inter-company relations (Korhonen, 2001).

J. Korhonen (2001) mentioned that the most commonly used industrial ecosystem analogy is for facilitating the emergence of industrial roundput systems, where the actors involved cooperate through waste material and energy utilization. He proposed to build industrial ecosystems considering natural ecosystem principles such as: roundput, diversity, locality, and gradual change (Korhonen, 2001). Table 1 sums up the following application of these four ecosystem principles in industrial systems.

Table 1: Natural Ecosystem Principles in Industrial Ecosystems

Principle	Natural ecosystem	Industrial ecosystem
Roundput	Recycling of matter Cascading of energy	Recycling of matter Cascading of energy
Diversity	Biodiversity Diversity in species, organisms Diversity in interdependency and co-operation Diversity in information	Diversity in actors, in interdependency and co-operation Diversity in industrial input, output
Locality	Utilizing local resources Respecting the local natural limiting factors Local interdependency, co-operation	Utilizing local resources, wastes Respecting the local natural limiting factors Co-operation between local actors

Gradual change	Evolution using solar energy Evolution through reproduction Cyclical time, seasonal time Slow time rates in the development of system diversity	Using waste material and energy, renewable resources Gradual development of the system diversity

Source: Korhonen J. (2001). Four ecosystem principles for an industrial ecosystem / Journal of Cleaner Production 9 (2001) http://engineering.dartmouth.edu/~d30345d/courses/engs171/industrialecosystem.pdf

Roundput

J. Korhonen proposes to use the term 'roundput', which may serve to describe ability industrial system to mimic the utilization of waste in the natural ecosystem, as it suits the purpose of being opposite to the term 'throughput', which is commonly used to illustrate the linear nature of the material flows in an industrial system; from raw materials, to products, to wastes (Korhonen 2001). William Ashworth used the term 'aroundput' for similar purposes and described its operation on the example of a forest ecosystem, its cyclic flow of materials (Ashworth 1995).

The carbon–oxygen cycle of photosynthesis exemplifies the way in which ecosystems operate. "Plants consume carbon dioxide and produce oxygen as a waste. Animals, in turn, require oxygen for respiration, but produce carbon dioxide as a metabolic waste" (Ayres and Ayres 1996). Ecosystems tend to keep most vital elements within the system and benefit from waste as food. Ecosystem's energy is recycled through flowing in food chains with the only driver of the system being the input from the 'infinite' solar energy (Korhonen, 2001).

The industrial system is supplied by 'free energy' stock of fossil fuels that has (up to now) made it possible to proceed with an unlimited growth paradigm. The fossil fuel stocks have enabled the industrial system to proceed with a throughput paradigm, beginning with raw materials and ending with products and wastes (Korhonen, 2001). But following the UN "Industrial Development Report 2016" (UNIDO, 2017), we can notice note that limiting factor of economic development is changing today as economic development is limited by natural capital instead of human manufactured capital. Thus, nowadays, the limiting factors are fishing population instead of the fishing boats, forests rather than sawmills, the oil in the ground instead of the pumping capacity that is the limiting factor (UNIDO, 2017).

To learn from the ecosystem principle of roundput, when facilitating the emergence of an industrial ecosystem, is understood to mean recycling of materials, and cascading of energy between the actors of the system. Roundput should encourage increasing reliance on renewable resources, involve further use of waste materials (recycling of matter), waste energy (cascading of energy) and waste fuels (Korhonen 2001).

Diversity

Ecosystem survival and stability is based on diversity – diversity in species, in organisms, in interdependency, in 'cooperation' and in information. By allowing high flexibility and adaptability, the existence of diversity should be seen as a long-term survival strategy of ecosystems as a result of permanently changing environmental conditions (Korhonen, 2001). Also, when the environmental conditions are almost constant, but severe resource limits exist, the coral reefs or tropical rain forests for example, optimize their functions through diversity (Ring, 1997).

Analyzing the artificial industrial, technological, and economic systems, we can notice that they are also diverse. For example, we can consider the diversity of product structures and supply. However, when we analyze the system under one single denominator, i.e. economic profit, the diversity is reduced. In addition, the diversity is reduced through the ideal of mass production, focusing on maximizing the rapid growth of homogenized industrial output products. However, the dominant Taylorist or Fordist ideal is beginning to be problematic by the rising emphasis on quality, variety, and diversity in the industrial output structure of products (Ayres and Ayres, 1996). Societal information regarding the ecosystem functions is in many cases based on parameters that are transformed to policy decision making by economic profit. It is very difficult to find monetary substitutes for natural goods. Therefore, human interventions into the natural ecosystem can be based on incomplete information and tend to be specialized on our needs without taking into consideration the existing diversity in nature (Korhonen, 2001).

The ecosystem principle of diversity, when considered in industrial environmental policy and management, could be used in cooperation diversity. The existence of diverse and interdependent cooperation systems could facilitate the existence of systems, where the actors involved use each other's waste materials and energy through cooperation. For example, the organic by-products of beer manufacturing can be used as food in stockbreeding, by-products of stock breeding can be used as fertilizers in growing grains for beer production. Actors that do not usually cooperate with each other would need to do so, e.g. large manufacturers, the public municipal organizations, waste management companies, consumers etc. (Korhonen, 2001).

As outcome of the diversity analogy, the diversity in industrial inputs and outputs would be promoted. The fuel basis of a power plant for example, could include the traditional coal and oil inputs but also peat, wood wastes, forestry wastes, recycled fuels from households or manufacturing. A case study of the regional energy supply system in Jyvaskyla region of Finland showed the alternative system based on cascading of energy with co-production of heat and power (CHP) principle (Korhonen 2001; Tekes 2006). In the heat and power co-production, the waste energy from electricity production can be used in the production of district heat for households and industrial steam for local industry requirements. The input basis of the Jyvaskyla energy supply system includes, to a large extent, forestry, saw-mill and plywood mill wastes. The technology of fluidized bed combusting is very flexible in respect to fuels as low-grade solid fuels like wood and saw mill wastes can be easily used as fuel. With the heat-and-power-co-production method, the output structure of the key power plant in the region is also comparatively diverse as electricity, heat and industrial steam can be produced from the same source (Korhonen, 2001).

Locality

Every species within the natural ecosystem adapt to the local environmental conditions and cooperate with its surroundings in diverse interdependent relationships. Ecosystems' components need to respect the local natural limiting factors (Wackernagel and Rees 1997). Regional economic or industrial systems usually are able to substitute the local natural limiting factors, of energy for instance, with imported fossil fuels. Furthermore, the assumption prevails that natural capital can be substituted with technological innovation or with human manufactured capital and therefore the sustainable use of regional or local resources has in many cases been mistreated (Korhonen, 2001).

To achieve the natural ecosystem's locality requirements, industrial systems would need to try and replace imported resources with local waste material and renewable energy sources. The regional industrial system would adapt to the local natural limiting factors. In such case, transportation would be reduced and cooperation with regional public and private actors would be enhanced, which will make us one step closer to inclusive and sustainable industrial development (Korhonen, 2001).

Gradual Change

Ecosystem survival relies on the energy flow from the sun and the evolution respects the renewal rate of the ecosystem. Ecosystem information moves through reproduction as evolution happens through the gene as the information storage medium. This process going in stark contrast compared to the cultural evolution, or its subset the industrial evolution (and how we used to call it – industrial revolution). In cultural and industrial evolution, culture serves as the information storage medium, e.g. oral, written, video, internet etc. records, and hence development happens much more rapidly than in the natural system (Norton, et.al., 1997). Here lies one of the fundamental problems of the environmental practice: a rapid increase in demand for a certain market product, for example, can lead to the extinction of the natural resource required for its production (Korhonen, 2001).

As we noted previously, the industrial evolution has been based on the non-renewable resources and fossil fuels, and therefore the renewal rate of the ecosystem was not respected. In sustainable industrial development, the gradual change principle could be interpreted as increasing the reliance on renewable flow resources not exceeding their renewal rate instead of using non-renewable stock resources. The industrial system should rely more on sustainable use of renewable natural resources as well as on the use of waste material and waste energy flows. In this way, industrial ecosystem adapts itself to the renewal time cycles of local natural resources instead of exceeding these with unlimited use of non-renewables that are often imported (Korhonen, 2001).

The forest industry of Finland could provide an example of a national industry, which is based on the renewable resource of the forest ecosystems. Annual cuttings are lower than the annual growth of the forests in Finland. In addition, the use of wood wastes, saw-mill wastes, forest residues, waste paper, black liquor from pulp mills, waste pulping chemicals and the implementation of the heat-and-power-co-production method in the industry power plants are factors that, when further developed, could make it possible for the whole forest industry to rely almost entirely on renewables or waste material and energy inputs (Tekes 2006). In other words, the application of the ecosystem principle of gradual change to an industrial system would encourage the use of renewable resources and waste material. In such way, energy and matter nature's time cycle could be respected in industrial systems (Korhonen, 2001).

Gradual change principle can be used in the industrial ecology approach to facilitate a gradual development of system diversity in a potential industrial ecosystem. Regional industrial systems are diverse systems, where ecological, economic, social, and cultural issues need to be considered when designing or trying to enhance the system interdependency and cooperation, for example, in waste utilization between the actors involved. Creating an industrial ecosystem 'from scratch' is very difficult in nowadays conditions; therefore, the direction to follow in industrial ecology could be to build on existing strengths. For example, a big industrial enterprise that can utilize a diverse fuel basis and this also promotes the other actors in the region to provide the actor with wastes to be used as fuels (Korhonen, 2001).

An environmentally active local governmental/municipal organization could provide the needed support for enhancing cooperation between private industrial actors that might otherwise be reluctant to engage in cooperation with each other because of competitive attitudes. Such process could be seen as a support system around which the industrial ecosystem development could be build.

KEY CONCEPTS OF INDUSTRIAL ECOLOGY

The key concepts of Industrial Ecology are based on understanding that the way natural resources are used and managed has economic, social, and environmental consequences on human health, resource availability, ecosystem quality, trade, and market prices or productivity. Achieving sustainable resource use and ensuring that the flows of materials are managed in an efficient way through the economic system is crucial, not only from an environmental perspective, but also from an economic and trade perspective (OECD, 2008).

Reid Lifset and Thomas E. Graedel (1992) proposed a framework for Industrial Ecology concepts which shows the study as operating at a variety of levels and can integrate them into a lager whole (Figure 6).

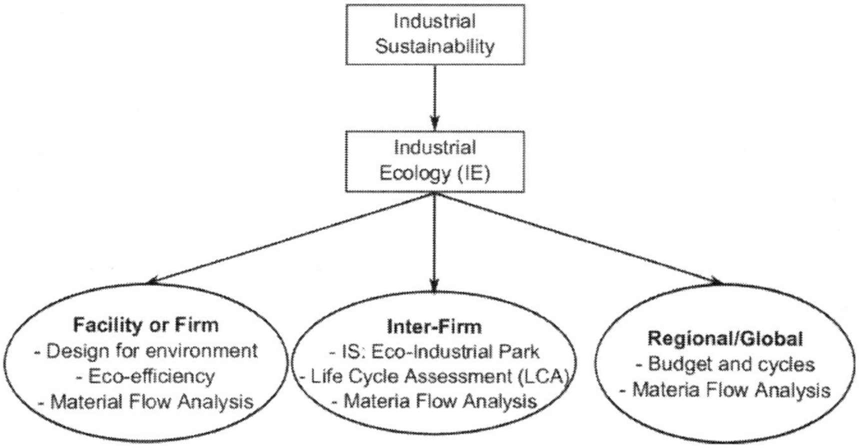

Figure 6: Concepts of Industrial Ecology Seen as Operating at Different Levels.

Source: Lifset, R., Graedel, T.E., (1992). Industrial Ecology: Goals and Definitions. Available at http://planet.botany.uwc.ac.za/nisl/ESS/Documents/Industrial_Ecology_Overview.pdf

Scientists divided Industrial Ecology concepts and tools between three levels (Lifset and Graedel, 1992):

- Firm or unit level;
- Inter-firm, district, or sector level;

- Regional, national, or global level.

While the firm and unit process is important, much of industrial ecology focuses at the inter-firm and inter-facility level, in part, because a system perspective emphasizes unexpected outcomes – and possibly environmental gains – to be related when a broader scope is used and because pollution prevention, a related endeavor, has already effectively addressed many of the important issues at the firm, facility or unit process level (Lifset and Graedel, 1992).

Andy Garner and Gregory Keoleian in their book "Industrial Ecology: An Introduction" pointed out 5 key concepts of Industrial ecology (Garner and Keoleian, 1995):

- Systems analysis;
- Material and energy flows and transformations;
- Multidisciplinary approach;
- Analogies to natural systems;
- Open- vs. Closed-loop systems.

Faye Duchina and Edgar Hertwichb (2003) stresses that many of the concerns and methods of Industrial Ecology overlap with problems and methods of Ecological Economics. The fields of Industrial Ecology and Ecological Economics are both issue oriented, and both take a systems approach to understanding and resolving challenges to the physical environment. There is a substantial intersection of concerns between the fields and, not surprisingly, an intersection also in their membership. Both fields are concerned with flows of substances, resources, energy, water, and wastes through an economy and with footprints, rucksacks, and other universal indicators of environmental health (Duchina and Hertwichb, 2003).

While the micro level of Industrial Ecology is comprised of physical balances for a growing number of materials and spatial units, the macro level is concerned with the formulation and evaluation of options for key decision makers. As in many other fields, there are substantial challenges in achieving conceptual and operational linkages between the micro and macro levels. An attempt to bridge this gap takes the form of an intermediate, or meso, level. This level represents analysis of extensive bodies of work in life cycle analysis and input-output economics and, in particular, by the growing linkages between them (Duchina and Hertwichb, 2003). That is why the International Society for Ecological Economics proposes 3-key concept for industrial ecological analysis:

- Material Flows, Energy Flows, and Environmental Accounting;
- The Product Level: Life Cycle Analysis;
- The Meso-Level: Input-Output Economics (Duchina and Hertwichb, 2003).

The International Society for Industrial Ecology describes a cluster of concepts and tools as hallmarks of industrial ecology:

- Life cycle assessment;
- Product-oriented environmental management and policy;
- Material and energy flow analysis (MEFA, also called socio-economic metabolism);
- Industrial symbiosis;
- Eco-efficiency;
- Dematerialization and decarbonization (ISIE, 2015).

Let's look at these concepts closer to understand how they can help us to reach the Industrial Ecology goals.

Systems Analysis

Industrial Ecology uses holistic (system) approach in analysis of the relationship between human activities and environmental problems. Framing the interaction between industrial and ecological systems, we should exploit a higher order systems tactic (Garner and Keoleian, 1995). One of the central key concepts of Industrial Ecology is the view that social and technological systems are bounded within the biosphere, and do not exist outside of it (Figure 7).

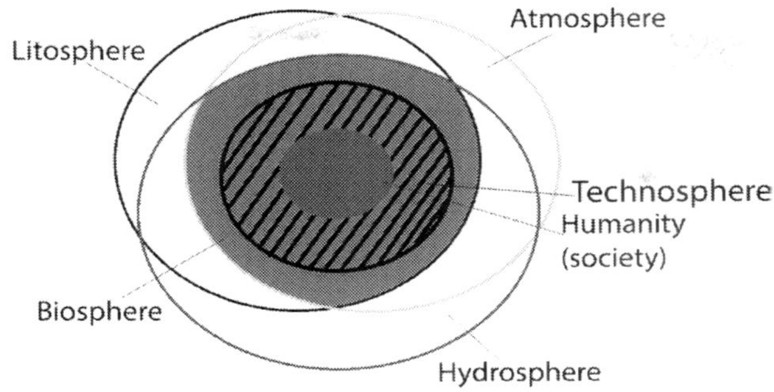

Figure 7: Position of Humanity (Society) and Industrial Systems (Technosphere) in the Global Ecosystem.

In many cases, the systems Industrial Ecology deals with are complex systems. Complexity makes it difficult to understand the behavior of the system and may lead to rebound effects. Due to unforeseen behavioral change of users or consumers, a measure taken to improve environmental performance does not lead to any improvement or may even worsen the situation (Garner and Keoleian, 1995).

Industrial ecology examines social issues and their relationship with both technical systems and the natural environment. Through this holistic view, Industrial Ecology recognizes that solving problems must involve understanding the connections, which exist between these systems, manufacturing processes cannot be viewed in isolation. Often changes in one part of the overall system can propagate and cause changes in another part. Thus, we can only understand a problem if we look at its parts through the net of interrelations in the whole ecosystem. Based on this framework, Industrial Ecology looks at environmental issues with a system thinking approach (Garner and Keoleian, 1995).

In system analysis, there are various system levels that may be chosen as the focus of study. For example, when focusing at the product system level, it is important to examine relationships to higher-level corporate or institutional systems as well as at lower levels, such as the individual product life cycle stages. One could also look at how the product system affects various ecological systems ranging from entire ecosystems to individual organisms. A system view enables manufacturers to develop products in a sustainable fashion (Garner and Keoleian, 1995).

The inherent recognition of the interrelationships between industrial and natural systems is central to the systems approach. In using systems analysis, one must be careful to avoid the pitfall that Kenneth Boulding has described: “seeking to establish a single, self-contained ‘general theory of practically everything’ which will replace all the special theories of particular disciplines. Such a theory would be almost without content, for we always pay for generality by sacrificing content, and all we can say about practically everything is almost nothing” (Boulding, 1956). We can surely say the same about Industrial Ecology study: if the scope of the research is too broad, the results become less meaningful; if too narrow – they may be less useful. A good examples of global scale system analysis of environmental problems are Meadows et al.’s Limits to Growth (New York: Signet, 1972) and Beyond the Limits (Post Mills, VT: Chelsea Green, 1992).

Urmila Diwekar (2005), talking about green process design, presents a systems analysis perspective that extends the traditional process design framework to green process design and industrial ecology leading to sustainability. For green process design this includes starting the design decisions as early as chemical and material selection stage on one end, and managing and planning decisions at the other end. However, doubts as well as multiple and conflicting objectives are inherent in such a design process. Uncertainties increase further in industrial ecology with the scale and duration of research. Combining approaches of industrial ecology with optimization methods, and constraining the system using thermodynamics and physics can provide a way to address this problem. The concept of overall sustainability goes beyond industrial ecology and brings in time dependent nature of the ecosystem and multi-disciplinary decision making. Optimal control methodologies and theories from financial studies can be useful in handling the time dependent uncertainties in this problem (Diwekar, 2005).

Material and Energy Flows and Transformations

A primary concept of industrial ecology is the study of material and energy flows and their transformation into products, byproducts, and wastes throughout industrial systems (Garner and Keoleian, 1995). The consumption of resources is accompanied along with releases of by-products to the air, water, land, and biota. The idea of industrial ecology is to build an industrial system with a full recycling and reuse of materials and minimal waste flows, which both reduces needs of primary materials and emissions to the environment. In this way, the main component of research in industrial ecology is the identification, quantification, and evaluation of resource and energy flows through industrial systems and their transformation in the environment (O'Rourke, et.al, 1996).

Organization of Economic Co-operation and Development (2008) defines the Material Flow Analysis (MFA) as "the study of physical flows of natural resources and materials into, through and out of a given system (usually the economy). It is based on methodically organized accounts in physical units, and uses the principle of mass balancing to analyze the relationships between material flows (including energy), human activities (including economic and trade developments) and environmental changes" (OECD, 2008).

Material flow analysis incorporates different methods: industrial and urban metabolism, input-output analysis, economy-wide material flow accounting, socioeconomic metabolism, and more recently material flow cost accounting (Moreau and Massard, 2017). Each method has specific scales, reference elements, and indicators. A material flow analysis study usually consists of a total of four consecutive steps on which consensus is found in the literature (OECD, 2010; Moreau and Massard, 2017). Process of material flow analysis consists of several stages, described below:

1. *Definition of the system.* A study begins with a definition of the problem and adequate goals. The choices are based on the most representative or critical element(s) or material(s) and on the classification of the system in space and time, the so-called system boundaries. Then the processes describing the relevant transformation, transport, and deposition paths within the system are defined. Finally, flows between the processes are identified (Moreau and Massard, 2017).
2. *Data acquisition.* During this stage, substance or material stock and flow data are collected from secondary sources, field studies or empirical surveys (Moreau and Massard, 2017).

3. *Calculation.* The quantification and overview of the stocks and flows should be completed on this stage. The complete material flow system is calculated and uncertainties are evaluated (Moreau and Massard, 2017).
4. *Representation and interpretation.* The results are represented in an appropriate way to allow visualization and ease of understanding and interpretation of the system. Sankey diagrams are commonly used. Conclusions focus on goal-oriented decisions based on the definition of the problem (Moreau and Massard, 2017).

The number of steps on every flow analyzing procedure can vary depending on research group, since no uniform standard exists, but the general procedure remains similar. The law of conservation of mass underlies every application, which implies that all material flows, as well as stocks, must be accounted for (Moreau and Massard, 2017). Figures 8 shows us an example of procedure for material flow analysis, proposed by P. H. Brunner and H. Rechberger (2004). They underlined the importance of iterative optimization.

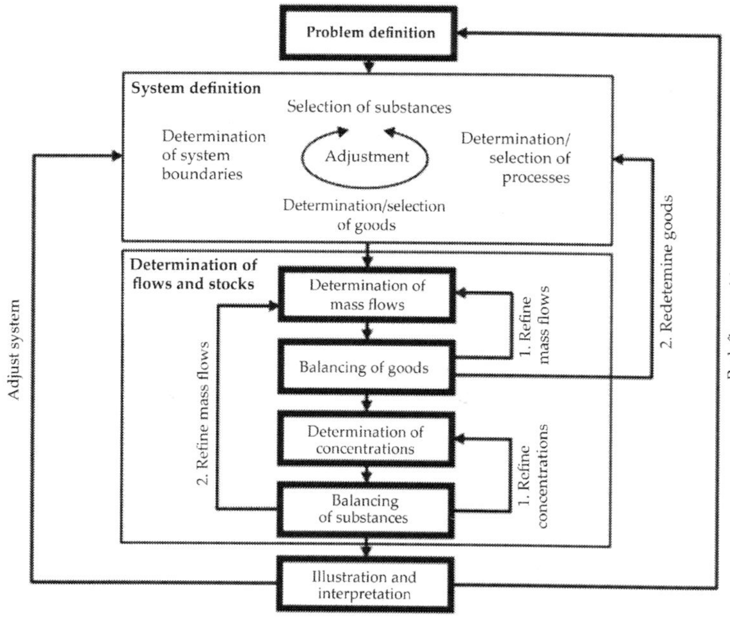

Figure 8: General Procedure for Material Flow Analysis.

Source: Brunner, P. H., Rechberger, H. (2004). Practical handbook for material flow analysis. Boca Raton, FL: CRC

The selections and provisions that are taken in the process of the MFA should be checked continuously. If necessary, they must be adapted and optimized to reflect the objectives of the project. The critical role of data in establishing reliable MFA has driven research on how to account for data quality and uncertainty in the analysis (Laner et. al., 2015). Scenario and sensitivity analyses are the most common ways to address uncertainties related to the different scales and diversity of data sources and the multiple scales of application in MFA (Moreau and Massard, 2017).

One strategy of industrial ecology is to lessen the amount of waste material and waste energy that is produced and that leaves the industrial system, subsequently impacting ecological systems adversely (Demacon, 2009). Figure 9 represents a schematic visualization of this strategy.

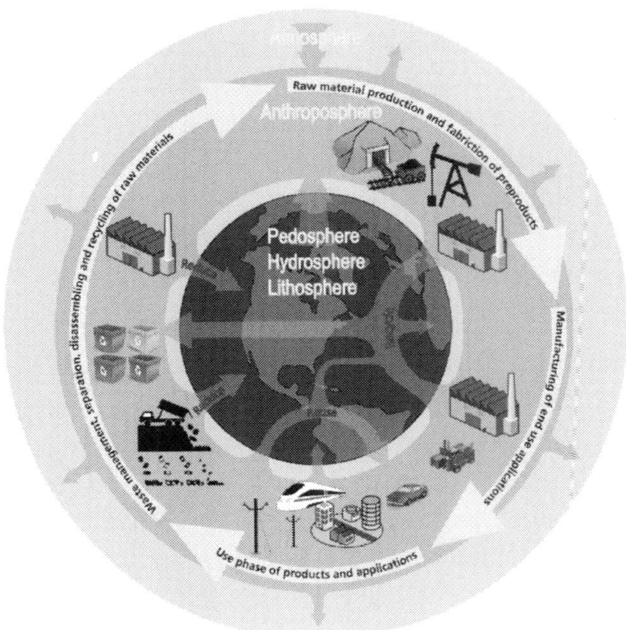

Figure 9: Analysis of Wastes Flows in Global Ecosystem.

Source: R-cubes Research Group, (2014). Industrial Ecology. Some basic definitions http://www.r-cubed-research.eu/Industrial_Ecology.htm

As we can see at Figure 9, efforts to utilize waste as a material input or energy source for some other entity within the industrial system can potentially improve the overall efficiency of the industrial system and reduce negative environmental impacts (R-cubes Research Group, 2014).

To identify areas to target for reduction, one must understand the dissipation of materials and energy (in the form of pollutants) – how these flows intersect, interact, and affect natural systems. Distinguishing between natural material and energy flows and anthropogenic flows can be useful in identifying the scope of human-brought effects and changes. Industrial ecology seeks to transform industrial activities into a more closed system by decreasing the dissipation or dispersion of materials from anthropogenic sources, in the form of emissions or waste materials, into natural systems. In the automobile operation example, it is useful to further trace what happens to the raw materials at the end of the products' lives in order to mitigate possible hazardous environmental impacts (Garner and Keoleian, 1995).

An important tool of industrial ecology, according to Dr. Hilko Van Der Voet, is the Substance Flow Analysis (SFA): "SFA aims to provide relevant information for an overall management strategy with regard to one specific substance or a limited group of substances. In order to do this, a quantified relationship between the economy and the environment of a geographically demarcated system is established by quantifying the pathways of a substance or group of substances in, out and through that system" (Van Der Voet, 2006).

We can use SFA as a specific case of the more broadly defined Material Flow Analysis. Material flow analysis concentrates on a particular system in general, e.g. a geographical region or an industrial process, and tries to track and quantify the resource flows both within the system and between the system and the environment. SFA, on the other hand, focuses on a specific substance/element and its way through an industrial process, respectively a value chain (Van Der Voet, 2006). Figure 10 shows the phosphorus flow analysis as an example of SFA on a European national level.

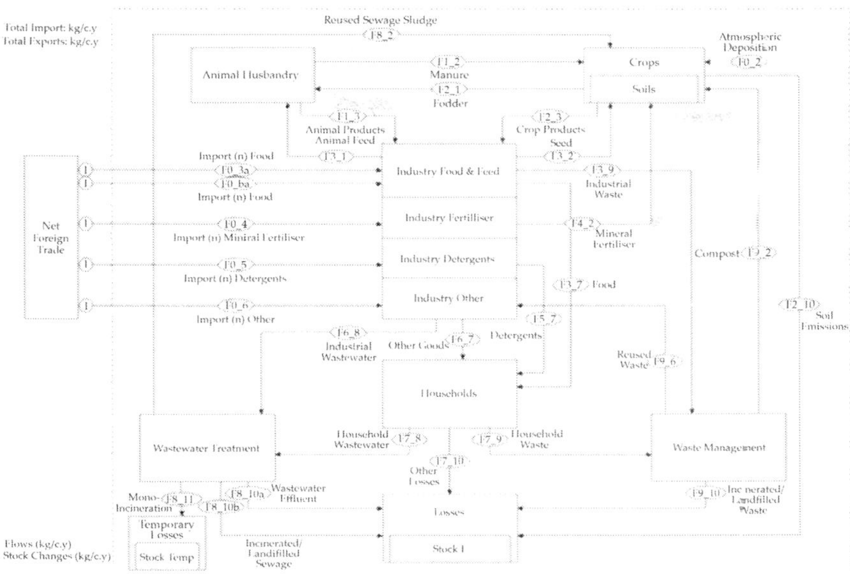

Figure 10: Analysis of P Stocks and Flows on the National Level.

Source: Jedelhauser, M., & Binder, C. R. (2015). Losses and efficiencies of phosphorus on a national level: A comparison of European substance flow analyses. Losses and Efficiencies in Phosphorus Management.

According to the principle that input equals output, a closed mass balance is accomplished in conclusion. Moreover, not only flows are taking into the account. Identifying material accumulations and their quantification is significantly important, especially in the case of toxic or radioactive substances (Van Der Voet, 2006).

Besides the environmental aspect, another – more recent – trend has been paying attention to the SFA methodology: The increased human demand for resources has led to a stronger competition on global raw material markets. In the case of energy resources this has already been observed for some time period, whereas for other resources, particularly metals, this issue is new. Here the SFA approach offers a contribution to an improved material utilization by making the metal flows and anthropogenic stocks more transparent, which is essential for minimizing the losses and designing the tailor-made recycling methods. Moreover, a model-based SFA is able not only to provide calculations but also to evaluate the effects of political instruments on material cycles through scenario simulations and analyses, such as the impact of a mandatory recycling quota on primary material demand (R-cubes Research Group, 2014).

Industrial Metabolism

'Metabolism' is another generic term widely used to characterize material accounting methods in Industrial Ecology. The word 'metabolism' comes from the Greek language, meaning 'to change' or 'to transform'. The concept of metabolism in biology describes the complex effect of numerous biochemical reactions that convert the substances in a living cell to support its growth and maintenance (Glasmeier, 2001).

The 'father' of Industrial Metabolism, Robert U. Ayres, wrote: "We may think of both the biosphere and the industrial economy as systems for the transformation of materials. The biosphere as it now exists is very nearly a perfect system for recycling materials. This was not the case when life on earth began. The industrial system of today resembles the earliest stage of biological evolution, when the most primitive living organisms obtained their energy from a stock of organic molecules ac-cumulated during prebiotic times. It is increasingly urgent for us to learn from the biosphere and modify our industrial metabolism, the energy- and value-yielding process essential to economic development. Modifications are needed both to increase reliance on regenerative (or sustainable) processes and to increase efficiency both in production and in the use of by-products." (Ayres, 1989).

The word 'metabolism', as used in its original biological context, implies the internal processes of a living organism. The organism consumes energy-rich, low-entropy materials (food), to provide for its own maintenance and functions, as well as a surplus to allow growth or reproduction. The process also necessarily involves excretion or exhalation of vital activity products, consisting of degraded, high-entropy materials. There is a captivating analogy between biological organisms and industrial activities, not only because both are materials-processing systems driven by a flow of free energy (Georgescu-Roegen, 1971), but because both are examples of self-organizing 'dissipative systems' in steady conditions, far from thermodynamic equilibrium (Ayres, 1988). The term 'industrial metabolism' derives from the notion that industrial economies, like biological organisms or natural ecosystems, can be understood as systems for material transformation with distinct metabolic pathways that evolve over time. Industrial metabolism offers a beneficial perspective for the study of collective human behavior and its effect on the materials cycles that comprise the workings of the natural environment. Social studies have considered industrial production and resource-use statistics as social indicators since the beginning of the Industrial Revolution. In the decades since the 1960s, industrial metabolism has been used as a paradigm to describe the exchange of materials and energy among industrial operations in a way comparable to the description of material and energy balances in natural ecological systems. This metabolic analysis contributes to the invention of technological, economic, and policy instruments with the main goal to find social-environmental balance (Glasmeier, 2001).

Ayres (1994) described, that at the most abstract level of design, the metabolism of industry is the whole integrated system of physical processes that convert raw materials and energy, along with labor, into finished products and wastes in a (more or less) steady-state condition. The manufacturing (supply) side of industrial system, by itself, is not self-managing. The stabilizing controls of the system are provided by its human component. This human role has two aspects:

- direct, as labor input;
- indirect, as consumer of output (i.e., determinant of final demand) (Ayres, 1994).

The system is stabilized, at least in its decentralized competitive market form, by balancing supply of and demand for both products and labor through the cost mechanism. In such way we can prove that the industrial economic system is, in essence, the metabolic regulatory mechanism (Ayres, 1994). "Industrial metabolism can be identified and described at a number of levels below the broadest and most encompassing global one. Thus, the concept is obviously applicable to nations or regions, especially 'natural' regions such as watersheds or islands. The key to regional analysis is the existence of a well-defined geographical border or boundary across which physical flows of materials and energy can be monitored" (Ayres, 1994).

Emphasis in metabolism study is concentrated on following the 'material flow' from its origins (e.g., mined ore) through the industrial processes used for its transformation, to the products that contain it, and finally to its disposal or re-entry into the industrial system. At its most wide-range, industrial metabolism links to human and ecosystem metabolism, for example, by considering the life cycle of metals and organic chemicals in plant tissue and the human diet. The information is used to construct a 'materials balance' of industrial system and make conclusions regarding the actions needed to improve the environmental character of its metabolism (Glasmeier, 2001).

One of important ways, in which the analogy between biological metabolism and industrial metabolism is useful, is concentration on the 'life cycle' of individual 'nutrients'. The hydrological cycle, the carbon cycle, and the nitrogen cycle are familiar concepts to earth scientists. The major way in which the industrial metabolic system differs from the natural metabolism of the earth is that the natural cycles (of water, carbon/oxygen, nitrogen, sulfur, etc.) are closed, whereas usual industrial cycles are open (Ayres, 1994). Robert Ayres proposed the comparison of natural and industrial material cycles in the form of four-box models (Figures 11-12).

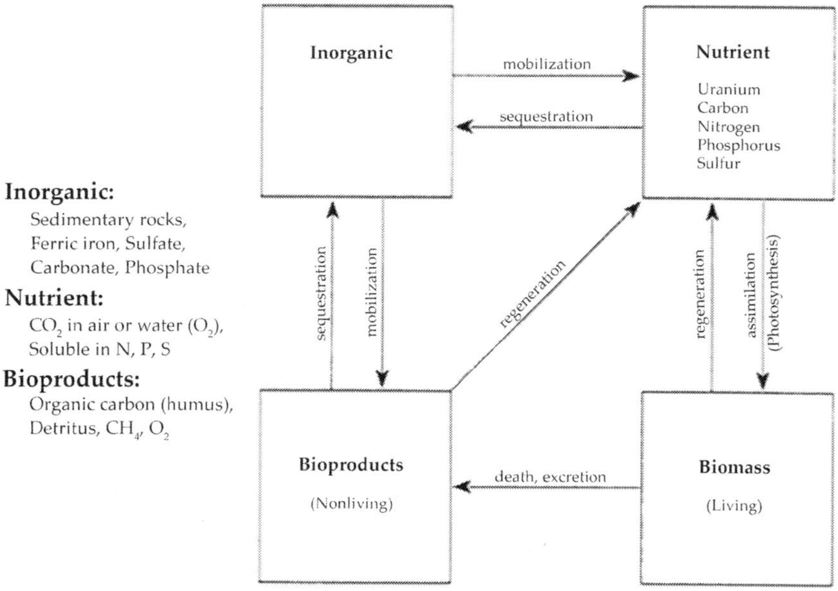

Figure 11: Box Scheme for Bio-Geo-Chemical Cycles.

Source: Ayres, 1994

The materials cycle (Figure 11), in general, can be visualized in terms of a system of sections containing stocks of one or more nutrients, linked by certain flows. For instance, in the case of hydrological cycle, the glaciers, the oceans, the freshwater lakes, and the groundwater are stocks, whereas rainfall, streams, and rivers are flows. A system is closed if there are no outside sources or sinks. In this sense, the Earth as a whole is essentially a closed system, except for the sun energy or occasional meteorite. A closed system becomes a closed cycle if the system is also in stable-state, that could be achieved when stocks in each compartment are constant and unchanging, at least on the average. The materials balance condition implies that the material inputs to each section must be exactly balanced (on the average) by the outputs (Ayres, 1994).

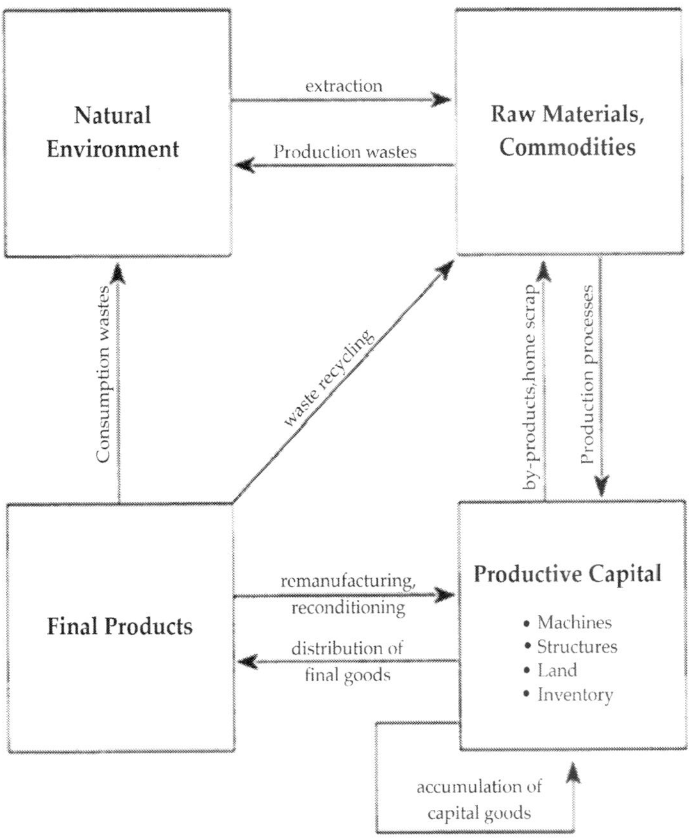

Figure 12: Box Scheme for Industrial Material Cycles.

Source: Ayres, 1994

As we see from the Figure 12, the industrial system does not generally recycle its nutrients. Rather, the industrial system starts with high-quality materials (fossil fuels, ores) extracted from the earth, and returns them to nature in degraded form (Ayres, 1994).

Ayres (1994) argued, that there are only two possible long-run fates for waste materials: recycling and reuse or dissipative loss (this is a straightforward implication of the law of conservation of mass.) The more materials are recycled, the less will be dissipated into the environment, and vice versa. Dissipative losses must be made up by replacement from virgin sources. A strong implication of the analysis sketched above is that a long-term (sustainable) steady-state industrial economy would necessarily be characterized by near-total recycling of intrinsically toxic or hazardous materials, as well as a significant degree of recycling of plastics, paper, and other materials whose disposal constitutes an environmental problem. Heavy metals are among the materials that should have to be almost totally recycled to satisfy the sustainability criterion. The segmentation of current metal supply needed to replace dissipative losses (i.e., production from virgin ores needed to maintain a stable level of consumption) is thus a useful surrogate measure of 'distance' from a stable-state condition, that is, a condition of long-term sustainability (Ayres, 1994).

To determine present and potential metabolic trails, industrial metabolism should consider the social dynamics that influence the integrity of the natural environment. Industrial metabolism associates indices of industrial production and social consumption to social and demographic indicators such as population, gender, age, land use, and economic activity. In such way, Industrial metabolism establishes quantitative relationships between material flows and social dynamics. To help in designing the public policy, industrial metabolism bases its findings in the context of government institutions, economic incentives, and cultural biases that influence the flow of physical resources and their impact on the natural environment processes and ecological laws. Possible indicated actions should include industrial process change, product substitution, broad public education, government regulation, development of enabling technologies, installation of new resource recovery capacity, and economic motivations by government (Glasmeier, 2001).

Ayres also described measures of Industrial Metabolism (Ayres, 1994): “A good measure of unsustainability is dissipative use. This raises the distinction between inherently dissipative uses and uses for which the material could be recycled or reused, in principle, but is not. The latter could be termed potentially recyclable”. He pointed out three important classes of materials use:

- material use that is economically and technologically compatible with recycling under present prices and regulations;
- material use that is not economically compatible with recycling but where recycling is technically possible, for example, if the separate waste collection problem were solved;
- material use for which recycling is inherently not feasible (Ayres, 1994).

As we can analyze, there is some indistinctness in this classification, but it should be possible for a group of international experts to arrive at some compromise. Generally speaking, not everybody will agree that most structural metals and catalytic agent are in the first category; other structural and packaging materials, as well as most freezing agents and solvents, fall into the second category. This leaves coatings, pigments, pesticides, herbicides, germicides, preservatives, flocculants, antifreezes, explosives, propellants, fire retardants, reagents, detergents, fertilizers, fuels, lubricants, and others in the third category. In fact, it appears that most chemical products belong in the third category, except substances physically embodied in plastics, synthetic rubber, or synthetic fibers (Ayres, 1994).

Transposed from the field of biology to describe human economic and management systems, ‘metabolism’ refers to variable dimensional levels, such as urban areas (urban metabolism); regions (regional metabolism); economic sectors (industrial metabolism); or more generally for any socioeconomic system (socioeconomic metabolism). In addition, most studies take a quasi-stationary approach, meaning that the network of procedures is interpreted into a set of linear equations describing how the flows of energy and materials partition among them. During analysis, the rate of accumulation is considered constant. When the temporal dimension is being simulated, the term ‘dynamic modeling’ is exploited and the main variables become functions of time (Moreau and Massard, 2017).

In the context of modern social and political awareness about environmental safety, the concept of industrial metabolism offers an analytical framework that is practically motivated to improve the quality of the natural environment through changes in industrial and social standards. Industrial metabolism examines the exchange of materials and energy between human society and nature at all spatial scales, applying the same analytical methods and proportions to describe material fluxes within and between natural and human systems. At the largest scale this includes study of anthropogenic contributions to atmospheric concentrations of trace gases and the flow of excess nutrients from agricultural activities to water bodies. At smaller scales, industrial metabolism addresses the use of resources in the human industrial activity through the development of resource-accounting frameworks for political and economic entities as well as life-cycle analyses of the materials used in the manufacture of industrial and consumer products (Glasmeier, 2001).

Multidisciplinary Approach

The next very important concept of Industrial Ecology is integrated approach or multidisciplinarity. Industrial Ecology takes into account aspects of three different disciplines: Technical Sciences, Environmental Sciences and Social Sciences (including Economics). Evan in his work of 1974 advocated multidisciplinary approach, discussed six aspects of Industrial Ecology:

- Technology;
- Environment;
- Natural resources;
- Biomedical aspects;
- Institutional and legal matters;
- Socioeconomic aspects (Evan, 1974).

The challenge is to merge them into a single approach. As we have analyzed in previous chapter, current definitions of Industrial Ecology rely heavily on engineered, technological solutions to environmental problems. Is still subject to debate how industrial ecology should balance the need for technological change with changes in consumer behavior. Some scientists see the study as having a narrow focused on industrial activity; to others – it is a way to view the entire global economic system.

Since Industrial Ecology is based on a holistic, systems view, it needs input and participation from many different disciplines. Furthermore, the complexity of most environmental problems requires expertise from a variety of fields – law, economics, business, public health, natural resources, ecology, engineering – to contribute to the development of industrial ecology and the resolution of environmental problems caused by industry. Along with the design and implementation of appropriate technologies, changes in public policy and law, as well as in individual behavior, will be necessary in order to rectify environmental impacts (Garner and Keoleian, 1995).

John Ehrenfeld takes a critical and very broad multidisciplinary perspective when claiming that to design a sustainable future, there is a need for other disciplinary frameworks than the positivistic and technocratic. There is a need for dueling paradigms - positivism and the natural science tradition on one hand – and historicism, hermeneutics and other variants all harking back to the phenomenology, on the other (Ehrenfeld, 1998).

Industrial Ecology Program at Norwegian University of Science and Technology (NUST) has the multidisciplinary approach that is emphasized by focusing on the theoretical, methodological, and experimental level both in social science as well as in technological issues (NUST, 2016). It is stressed that multidisciplinary approach to complex problems is an absolute necessity. Without this approach, it is not possible to analyze, far less understand, the connection and cause-effect relations in the intersection between technology, industry, market, consumer, governmental institutions, and the natural environment (Brattebo et. al., 1998).

Analogies to Natural Systems

There are several valuable analogies between industrial and natural ecosystems. The natural system has evolved over many millions of years from a linear (open) system to a cyclical (closed) system in which a dynamic equilibrium between organisms, plants, and the various biological, physical, and chemical processes in nature was created. In case of natural ecosystem, nothing leaves it, because wastes are used as substrates for other organisms (Garner and Keoleian, 1995).

Ehrenfeld (1998) argues that natural ecosystems, the source of the ecological metaphor in industrial ecology, offer the only example available to us of long-lived, robust, resilient living systems. These systems are the only kinds of system which offer "the three C's, connectedness, community, and co-operation, in many ways the exact opposite characteristics of what we find in the industrial community today" (Ehrenfeld, 1998). Thus, Industrial Ecology draws the analogy between industrial and natural systems and suggests that the main objective is to stimulate the evolution of the industrial system so that it shares the same characteristics as natural ecosystems. Our goal is to reach this dynamic equilibrium and high degree of interconnectedness and integration that exists in nature (Ehrenfeld, 1998).

James J.Kay described natural ecosystems as complex adaptive self-organizing hierarchical open (SOHO) systems. Figure 13 shows us the conceptual model for self-organizing systems as dissipative structures (Kay, 2002).

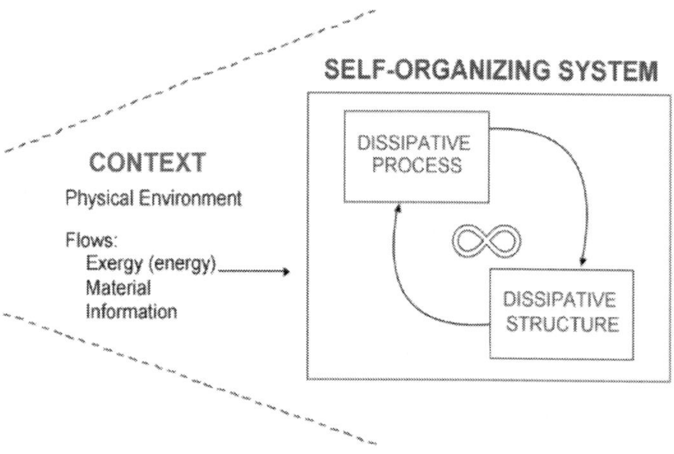

Figure 13: Conceptual Model for Self-Organizing Hierarchical Open (SOHO) Systems.

Source: Kay, J.J., (2002). On Complexity Theory, Exergy and Industrial Ecology: Some implication for Construction Ecology.

Spontaneous comprehensible behavior and organization occurs in open systems, such as natural ecosystems. Central to understanding such phenomena, according to J. Kay, is the recognition that open systems are processing and enduring flow of high quality energy (exergy). Self-organizing dissipative processes emerge whenever sufficient energy is available to support them. Dissipative processes rearrange the available raw materials in order to dissipate the energy. Through catalyze, the information enables and promotes some processes to the disadvantage of others. The physical environment will favor certain processes. The interplay of these factors defines the context for (i.e. constrains) the set of processes which may emerge (Kay, 2002).

After a dissipative process emerges and becomes established, it reveals itself as a structure. According to Kay (2002): "these structures provide a new context, nested within which new processes can emerge, which in turn beget new structures, nested within which.... Thus, emerges a SOHO system, a nested constellation of self-organizing dissipative process/structures organized about a particular set of sources of exergy, materials, and information, embedded in a physical environment". As we can understand, the canon of the SOHO system is the complex nested interchange and relationships of the processes and structures, and their propensities, that give growth to coherent self-perpetuating behaviors, that define the attractor (Kay, 2002).

High degree of integration and interconnectedness is a special feature of natural systems. They have a food web by which all organisms feed or are eaten as a food source by other members of the web, and are recycled as wastes. Natural ecosystems have a complex system of feedback mechanisms that induce reactions should certain limits be reached. Both natural and industrial systems have cycles of nutrients, resources, and energy. These cycles are integral to the functioning and equilibrium of the entire natural system; material and energy flows through various products and processes are integral to the functioning of the industrial system (Garner and Keoleian, 1995).

Ecological communities provide resources, energy, and information required for human societies to sustain themselves (Kay, 2002). This is depictured in Figure 14.

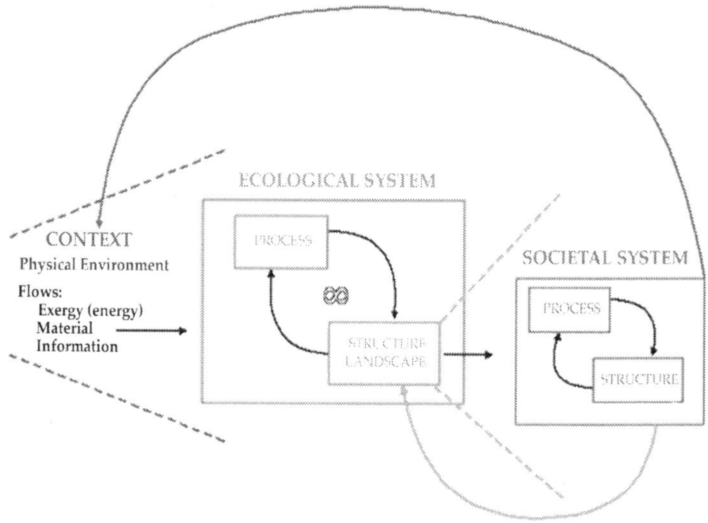

Figure 14: Conceptual Model of the Ecological-Societal System Interface – Single Horizontal Level.

Source: Kay, J.J., (2002). On Complexity Theory, Exergy and Industrial Ecology: Some implication for Construction Ecology.

The societal system, as Kay (2002) calls social system, depends on the flow of exergy, materials, and information from the ecological system to support its processes and structures. These flows, along with the biophysical environment provided by the ecosystems, are the context for societal systems. Or we can present this as the context constraining the possible social processes and structures in a specific territory. While Figure 14 illustrates a single ecological system providing the context for the societal system, the reality is that it is a suite of adjacent ecological systems (i.e., fields and wetlands adjacent to a farm) that provide this context. Alteration in these adjacent systems will damage the context and thus the possibilities for the systems in question (Kay, 2002).

Kay (2002) admits that societal system can also influence the ecological system in two ways. The first is the influence through changes in the structure of the ecosystem (i.e., cutting grass on the fields or filling in wetlands, and all the human activities which involve removing or dismantling ecological structures on the landscape). Such actions, of course, modify the flows from the ecological systems to the societal systems and thus create a feedback structure on the landscape. This is represented in Figure 14 by the lower arrow back from the social system to the structure landscape in the ecosystem. The feedback to the social system arises because changes in the ecological structure influence the context for the social system. The second influence occurs when the context of the ecological system is altered by the social system (i.e., the runoff into a wetland or stream may be transformed by human activities on adjacent properties). It is shown in Figure 14 by the upper arrow from the social system back up to the context of the physical environment. This influence is qualitatively different than the structural influence just described (Kay, 2002).

The resulting feedback loop has more steps and accordingly is more indirect. Analyzing these interrelationships, we can clearly see, that by changing the context of the ecosystem, humanity affects the natural processes and, in turn, the ecological structure and ultimately damage the social system's own context. The bright practical example of this theory is the burning issue of climate change on our planet. Humanity, burning non-renewable fossil fuels, altered the Earth's atmosphere and now we can feel the natural ecosystem's feedback in the form of severe weather, hurricanes, floods, drouths, and other natural disasters (Kay, 2002).

James Kay (2002) developed a set of design principles for Industrial Ecology, that can be used to create industrial ecosystem (to emerge the production-consumption system into the natural ecosystem):

- The interface between social systems and natural ecosystems reflects the limited ability of natural ecosystem to provide energy and absorb waste before their survival potential is significantly damaged, and that the survival potential of natural ecosystems must be maintained. This is referred to the principle of interfacing.
- The behavior and structure of large scale social systems should be as similar as possible to those displayed by natural ecosystems. This is referred to the principle of bionics (in the Industrial Ecology it is often referred as mimicry).
- Whenever feasible the function of a component of a social system should be carried out by a subsystem of the natural biosphere. This is referred to the principle of appropriate biotechnology.

- Non-renewable resources are used only as a capital expenditures to bring renewable resources on line (Kay, 2002).

Natural ecosystem principles discussed above, in combination with the vision of complex systems theory, provide a theoretical basis for an ecosystem approach to Industrial Ecology. Industrial Ecology is taken to be the activity of designing and managing human production-consumption systems, so that they interact with natural systems, to form an integrated anthropogenic ecosystem which has ecological integrity and provides humans with a sustainable livelihood. In essence, Industrial Ecology is about designing human industrial ecosystems which fit in with natural ecological systems (Kay, 2002).

Korhonen (2001) proposes his vision of ideal industrial ecosystem based on four principles, which we discussed in the Chapter 3.1. Perfect industrial ecosystem would be constructed from the two systems: the industrial subsystem and the mother ecosystem. The two systems would operate through the same principles of system development, i.e. roundput, diversity, locality, and gradual change. Together the systems would form a single system. (Figure 15).

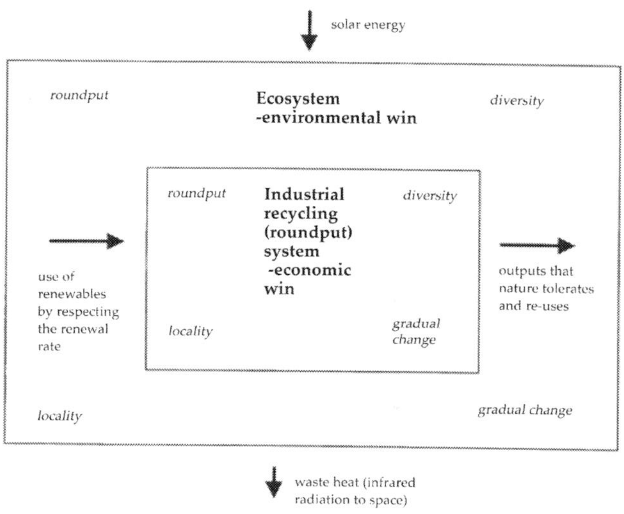

Figure 15: Vision of the Perfect Industrial Ecosystem.

Source: Korhonen, J., (2001). Journal of Cleaner Production 9 (2001)

Such a system would be constructed from the industrial subsystem and the mother ecosystem in which the industrial system is rooted. The only input to the system as a whole would be the 'infinite' solar energy and waste heat would be the only output from the system. The inputs to the industrial system would include only renewable resources and recycling of matter as cascading of energy flows would then happen between the industrial members within the system. All by-products that are dumped into the natural system from the industrial system would include only such things that nature can tolerate or perhaps re-use and recycle in its own operation (Korhonen, 2001).

In a perfect industrial ecosystem, both of the systems (the industrial subsystem and the mother ecosystem) operate according to the same principles of system functioning: roundput, diversity, locality and gradual change. In other words, the industrial system is also based on recycling, diverse interconnections, and cooperation. The industrial system would be arranged as a local cooperation system that respects the local natural limits and develops gradually ensuing the renewal rate or the reproduction cycle of the local renewable resources (Korhonen, 2001).

The vision presented by Korhonen (2001), is, of course, highly idealized and the industrial system will always differ from the ecosystem operation. But the goal of the vision is what the industrial ecology calls type III ecology (Allenby and Cooper,1994): "an ecosystem that has evolved from linear material and energy flows into a situation where the resources of life are limited and therefore the system operates through almost complete cyclic nature of the material flows". In type III ecology the only external resource is the solar energy; and 'resources' and 'wastes' are indeterminate as 'waste is a resource' (Korhonen, 2001).

The perfect industrial ecosystem would be an industry – environment 'win – win situation' (Porter and Van der Lindec, 1996). As in theory, when properly implemented, recycling can reduce the raw material and energy input to the industrial system as well as the waste and emission output from the system. In this case, the environmental 'win' would happen. On the other hand, the economic gains appear as raw material and energy costs are reduced if waste serves to substitute the need of these inputs. Similarly, costs environmental taxes can be reduced. Further, waste management costs as well as 'image costs' can be saved and possibly the green market potential can be better utilized (Korhonen, 2001). In this case, the industrial/economic 'win' would happen.

Industrial Symbiosis

Industrial Symbiosis as one of the key concepts of Industrial Ecology looks at the flows of resources that are interacting between companies and can be seen as a practical application of Industrial Ecology principles. An Industrial Metabolism is composed of materials and energy flows that are transferred through the system, while Industrial Symbiosis is broader and also looks at the relation with the biosphere (Garner and Keoleian, 1995).

Industrial Symbiosis is a local collaboration where public and private enterprises buy and sell local and residual products, resulting in mutual economic and environmental benefits (Kalundborg Symbiosis, 2015). "It is a feature of the industrial ecosystem thinking – establishing symbiotic relationships between separate industries in geographical proximity" (Lowe, 1997). From a supply chain perspective, Industrial symbiosis is an "industrial network including diverse and tight relationships between independent companies across supply chains" (Miemczyk et al. 2012).

The network of Industrial Symbiosis is based on information, energy and by-products flows between industries involved. Marian Chertow (2007) sets up three main opportunities giving incentive for resources sharing:

- The first opportunity is to reuse wastewater or by-products by means of infrastructures like a pipe, to be transported elsewhere to another company;
- The second opportunity is facilities/infrastructures sharing;
- The third opportunity is joint provision of services between the companies (Chertow, 2007).

Industrial symbiosis involves different independent industries in the exchange of by-products and residual resources, such as excess energy and water (Chertow, 2000). Consequently, supply chain symbiosis in industrial networks is the display of the flow of products, by-products, residual resources, and information. Gábor Herczeg et al. (2013) worked out the schematics of supply chain to help in development of industrial symbiosis systems (Figure 16).

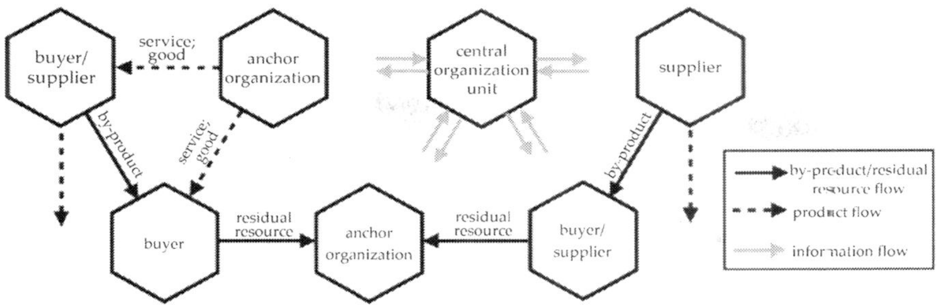

Figure 16: Schematic of Supply Chain Symbiosis in the Industrial Network.

Source: Proceedings of the 20th International EurOMA Conference, June 7-12, 2013, Dublin, Ireland, paper SCM-19 5

In an industrial network, symbiosis usually involves different types of organizational interdependencies across supply chain levels (see Figure 16). The structure of industrial symbiosis is risen up from:

- dyads, which are pairs of companies exchanging their by-products;
- star networks, where a sole producer of a good or service is directly connected to its customers;
- fully connected networks, in which all members have direct relations to each other (Ashton, 2008; Herczeg et al., 2013).

Dyadic relationships usually develop between manufacturers of cement, fertilizer, oil, biofuel, paper, pharmaceutical, food and mineral industries and involve farmers, fisheries, or breweries (Zhu and Cote, 2004; Duflou et al., 2012). Plants in industrial symbiosis often settle around so called anchor organizations such as water suppliers, waste treatment facilities and energy providers (Chertow, 2007). Star networks are typically built around the anchor organization with several plants sharing its goods and services (Ashton, 2008). We can conclude that industrial symbiosis often has a central organizational unit which is responsible for connecting all the actors, disseminating knowledge and ensuring central focus on sustainable development (Chertow, 2007).

In industrial symbiosis, demand for by-products is generated by the need of raw materials, which they can successfully substitute. In other words, a by-product or waste of one company is a feedstock for another company's manufacturing process. In general, in industrial symbiosis suppliers and buyers conduct market transactions similar to vertical supply chains as well as share resources and services in a horizontal manner. Supplying by-products often requires additional conversion, transportation, and substitution processes (Herczeg et al., 2013).

Operations in industrial symbiosis involve the obtaining, production, storage, and transportation of goods and by-products, as well as the concurrent use of services. At the same time, as it provides raw materials to buyers, the association in an industrial symbiosis offers the suppliers of the goods a treatment of residual resources that in traditional industrial system would be waste waters or solid wastes. Plants, in addition, also benefits from excess energy (e.g. heat, steam) obtained from power plants and refineries (Chertow, 2000).

Industrial ecosystems that are realized with a concept of Industrial Symbiosis are called Eco-parks or Eco-industrial parks. An eco-park is "an industrial ecosystem consisting of many companies (clusters), often located in the same area" (Ehrenfeld and Gertler 1997). The purpose with the parks is to realize better economic and environmental performances when companies are co-operating on logistic, waste exchange, energy exchange etc. The best-known example of an eco-park is Kalundborg, Denmark (Figure 17). Kalundborg Symbiosis is the world's first real working industrial symbiosis. The Kalundborg Symbiosis is an industrial ecosystem, where the by-product residuals of one enterprise are used as resources by other enterprises, in a closed cycle (Kalundborg Symbiosis, 2015).

Figure 17: Kalundborg Symbiosis, Denmark.

Source: Kalundborg Symbiosis, 2015. http://www.symbiosis.dk

Eco-park, presented at the Figure 17, represents an attempt to model an industrial park after an ecological system. There are 11 companies, including Statoil and Kemira, connected in cycles exchanging energy, by-products/resources and wastes. The companies in the park are highly integrated and utilize the waste products from one firm as an energy or raw material source for another as it is shown at the Figure 18 (Kalundborg Symbiosis, 2015).

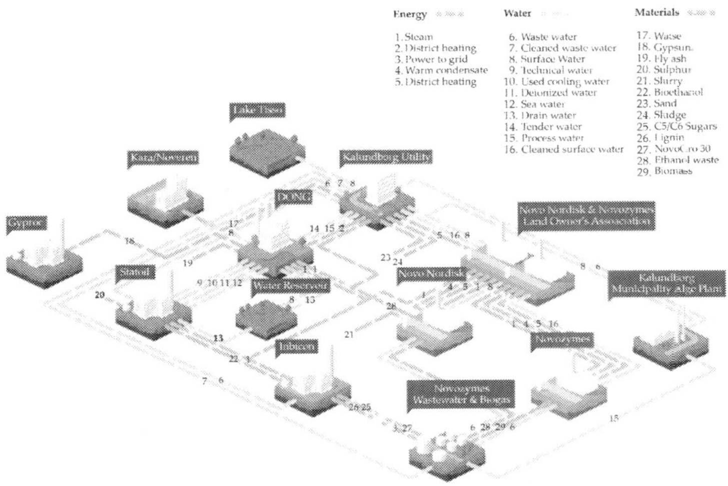

Figure 18: Diagram of Kalundborg Symbiosis.

Source: Kalundborg Symbiosis, 2015. http://www.symbiosis.dk/diagram

The Kalundborg Industrial Park was not originally planned for industrial symbiosis. Its current state of waste heat and materials cycling developed over a period of 40 years. The park began in 1959 with the startup of the Asnas Power Station. The first episode of sharing between two entities occurred in 1972 when Gyproc, a plaster-board manufacturing plant, established a pipeline to supply gas from Tidewater Oil Company. In 1981, the Kalundborg municipality completed a district heating distribution network within the city of Kalundborg, which utilized waste heat from the power plant. Early sharing at Kalundborg tended to involve the sale of waste products without significant preparation. Each further link in the system was negotiated as an independent business deal, and was approved only if it was expected to be economically beneficial.

Since then, the facilities in Kalundborg have been expanding, and have been sharing a variety of by-products and secondary resources, some for the industrial symbiosis needs and some out of necessity, for example, freshwater scarcity in the area has led to water reuse schemes (Kalundborg Symbiosis, 2015). Figure 19 shows chart of by-products and secondary resources exchange inside of Kalundborg Symbiosis system.

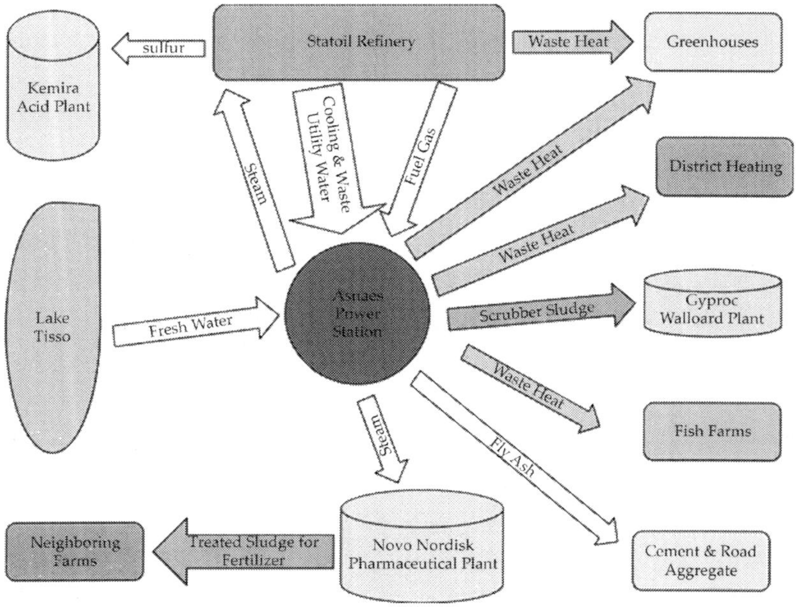

Figure 19: Flow Chart of Material Sharing Between Entities in the Kalundborg Eco-Industrial Park.

Source: Kalundborg Eco-Industrial Park Symbiosis. Wikipedia. https://en.wikipedia.org/wiki/Kalundborg_Eco-industrial_Park#/media/File:Kalundborg_Eco-Industrial_Park_Symbiosis_Map.jpgS

In the development of the Kalundborg Symbiosis, the most important elements are the healthy communication and the strong cooperation between the participants. The symbiosis was founded on human relationships, and fruitful collaboration between the employees that had made the development of the symbiosis-system possible. The 2010 results of this project include:

- 3 million m^3 of water saved through recycling and reuse;

- Biogas is obtained from yeast slurry as a result of insulin production;
- Recycling of 150,000 tons of gypsum from desulfurization of flue gas (SO_2) replaces import of natural gypsum ($CaSO_4$) (Kalundborg Symbiosis, 2015).

The idea of industrial symbiosis is to integrate companies in a framework of services, by-products and energy flows that would make them dependent on each other. This would lead to the creation of a flexible entity that performs as an ecosystem does, where the outputs of each one becomes the inputs of another. Therefore, it results in a cyclic system where there is no energy loss (Hawken, et.al., 2008) and where the concepts of 'resources' and 'waste' disappear. Implementation of industrial symbiosis concept would have results on a macroeconomic, microeconomic but also environmental level, as it tries to address such issues (Almasi et.al., 2011).

On macroeconomic perspective, industrial symbiosis is a response to environmental concerns regarding materials rarefaction. Such system allows – by re-using by-products, wastewater, or flow of outgoing energy - to minimize the need for extraction of raw materials and use of electricity. Industrial symbiosis also leads to reduce overall greenhouse gas emission, as the work paper of Steve Harris (2007) shows. By sharing services, infrastructures and avoiding landfills disposal, the Kwinana IS system (Australia) resulted in decreasing by 170.000 tons per annum its CO2 emission (Harris, 2007). On an economic basis, the formation of an interdependent members' network creates a diversified and flexible industrial area that adapts to changes, regarding environmental and legislative matters. It brings up a phenomenon of 'clusterization' between companies that leads to external economies of scale (Wahl, 2008).

As Almasi (2011) mentioned, these economies of scale result from positive externalities related to the size of the industrial area. They provide a broad range of opportunities to the manufacturers: a close and qualified employment pool, a common mobile labor force and the possibility of sharing services, infrastructures and, in such way, their costs. Moreover, the Network effect theory (Metcalfe's law) suggests that, as the number of companies is growing linear, the number of possible synergies grows in a geometrical way. The same relation is true for the economic utility of being in such system. This utility value will increase up to a point where new companies will decide to join in. This increase will feed the virtuous cycle of utility value creation. Industrial symbiosis combines economic growth and reduction of the industrial impacts on the environment (Almasi et.al., 2011).

On microeconomic perspective, looking at the companies on an individual level, the outcomes in terms of reduction of their impacts on the environment are obvious. Operating in the industrial symbiosis system, the company usually adjusts its production process. It uses outputs of another company and decreases its use of raw materials (Almasi et.al., 2011). For example, company uses waste of another factory, which previously was disposed in landfills polluting soil and groundwater. Moreover, the outputs generally circulate by means of fixed infrastructures that reduce the use of vehicles – that generate greenhouse gases – in order to dispose waste in landfills (Harris 2007).

On the energy level, industrial symbiosis allows synergies resting on the exchange of energy flows, thus avoiding a too strong loss related to a high entropy effect. Energy is produced only once and it is then reused. It reduces the dependency of the industrial system on fossil fuels and other non-renewable sources (Almasi et.al., 2011).

Of an economic outlook, the arguments are numerous. Companies that compose this system are very diverse. Although they adopt an individual approach of their own product markets, they remain collectively united and organized in their environment. It may end in providing them with competitive advantages on their respective markets. For example, by transferring its wastes to other manufacturers, the company avoids landfills disposal costs. Moreover, this exchange can be remunerated and contribute to a rise of the company income. It is a solution where everyone benefits. In most of the cases, a reorganization of the production process is necessary, but it does not necessarily imply additional costs (Almasi et.al., 2011). Taken as a complex web of interconnections, such a system is flexible and dynamic and could be used as a method for achieving the inclusive and sustainable industrial development.

Open- vs. Closed-loop Systems

The evolution of the industrial system from a linear system, where resources are consumed and damaging wastes are dissipated into the environment, to a more closed system, like that of ecological systems, is a central concept to industrial ecology (Garner and Keoleian, 1995).

Braden Allenby has described this change as the evolution from a Type I to a Type III system. Type I Linear System is shown at Figure 20.

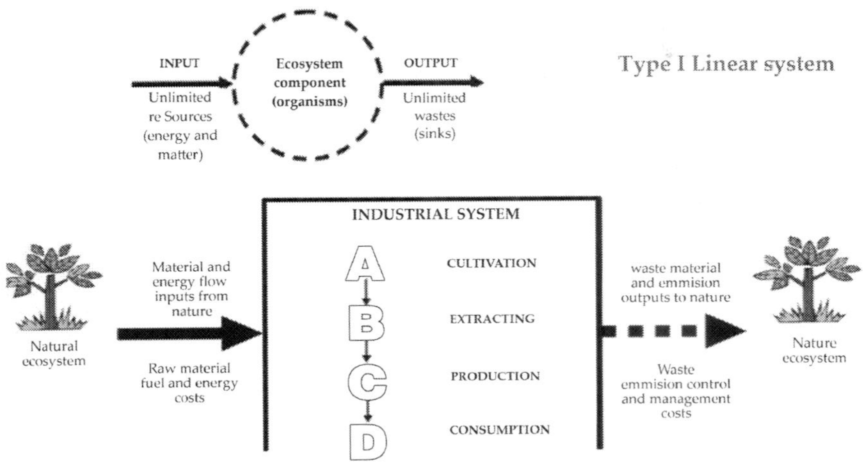

Figure 20: Type I Linear System.

Source: Denesia, M., (2017). SlideSare, Available at: https://www.slideshare.net/MerlynH/industrial-ecology-55184762

A Type I system is depicted as a linear process in which materials and energy enter one part of the system and then leave either as products or by-products/wastes. Because wastes and by-products are not recycled or reused, a large, constant supply of raw materials supports this system. Unless the supply of materials and energy is infinite, this system is unsustainable; further, the ability for natural systems to assimilate wastes (known as 'sinks') is also finite (Garner and Keoleian, 1995).

In a Type II system some wastes are recycled or reused within the system while others still leave it (Figure 21).

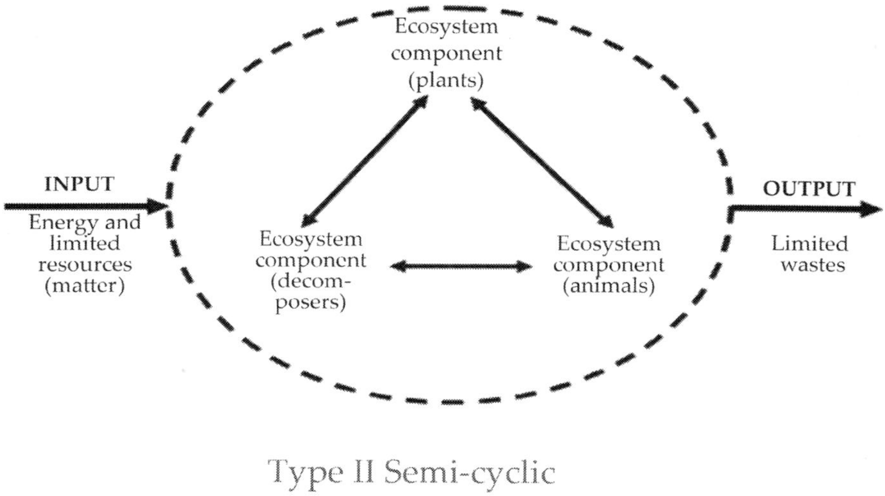

Figure 21: Type II Semi-Cyclic System.

Source: Denesia, M., (2017). SlideSare, Available at: https://www.slideshare.net/MerlynH/industrial-ecology-55184762

This type characterizes much of our present-day industrial systems. A major bulk of materials eventually contributes to the landfill disposal. Only a fraction of material is recycled indefinitely (Garner and Keoleian, 1995).

A Type III system represents the dynamic equilibrium of ecological systems, where energy and wastes are constantly recycled and reused by other organisms and processes within the system as it is shown at the Figure 22 (Garner and Keoleian, 1995).

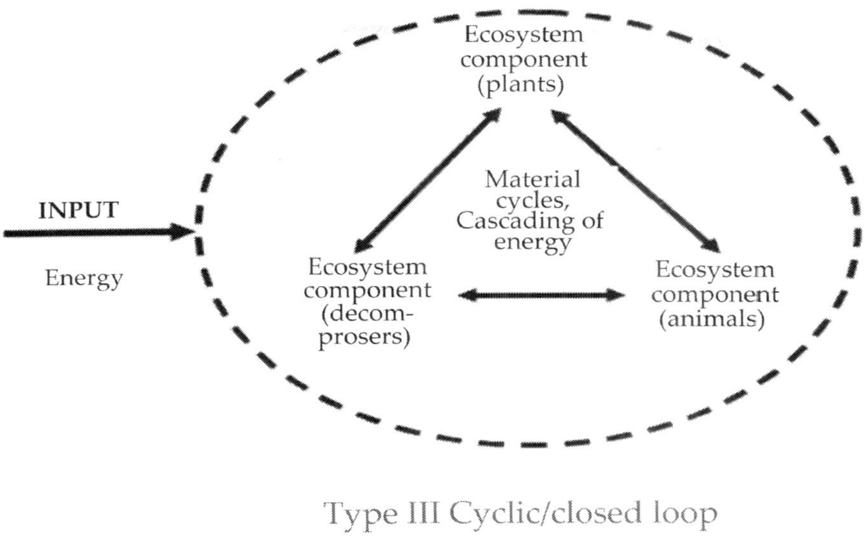

Figure 22: Type III Closed-Loop System.

Source: Denesia, M., (2017). SlideSare, Available at: https://www.slideshare.net/MerlynH/industrial-ecology-55184762

This is a highly integrated, closed system. In a totally closed industrial system, only solar energy would come from outside, while all byproducts would be constantly reused and recycled within. No permanent waste is generated – all materials are returned as nutrients for natural or technical food chains. A Type III system represents a sustainable state and is an ideal goal of industrial ecology (Garner and Keoleian, 1995).

Mark Fedkin developed open-loop / closed-loop concepts for material recycling. Open-loop recycling basically means that a material is not recycled indefinitely and is eventually excluded from the utilization loop and becomes waste. The diagram in Figure 23 shows a material flow through the linear (open-loop) system. In this representation, stocks are shown with rectangular boxes, and transforming processes are shown by hexagon boxes (Fedkin, 2017).

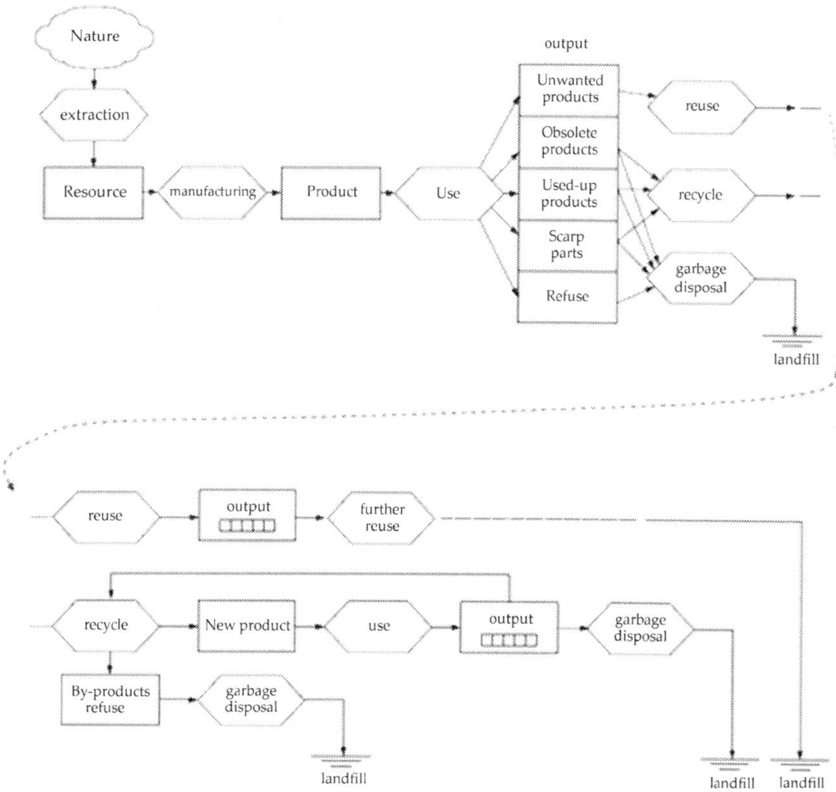

Figure 23: Open-Loop Material Flow ("Cradle-to-Grave").

Source: Fedkin, M., (2017). Recycling: open-loop versus closed-loop thinking.

At the figure above, we can see that natural resources extracted from the environment are transformed into a product via manufacturing process. After its use, the product may be discarded as one of the outputs:

- whole product that is not needed anymore,
- whole product that became outdated (although still functional),
- non-functional or old product because of its limited lifetime,
- recyclable / reusable parts or scrapped materials,
- non-recyclable waste (Fedkin, 2017).

As we see at the Figure 23, those outputs enter one of the post-use channels – reuse, recycle, and garbage disposal, the latter contributing to the landfill. Reuse channel is usually limited, just delaying garbage disposal. Recycling loop results in producing another material, which is typically has lower quality and purity than the original material. It may be transformed further into a different product, which after use creates similar outputs. Eventually, a small part of the original resource may be stuck in the loop, but the majority of it becomes disposed of. The essential point is: even if recycling and reuse are involved, eventual down-grading renders material non-usable, and it contributes to waste generation in the end of the lifecycle. Open-loop recycling postpones disposal and slows down extraction of new raw materials, but does not provide ultimate solution to the problem (Fedkin, 2017).

According to Fedkin (2017), closed-loop recycling is a more sustainable concept, which means that recycling of a material can be done indefinitely without degradation of properties (Figure 24).

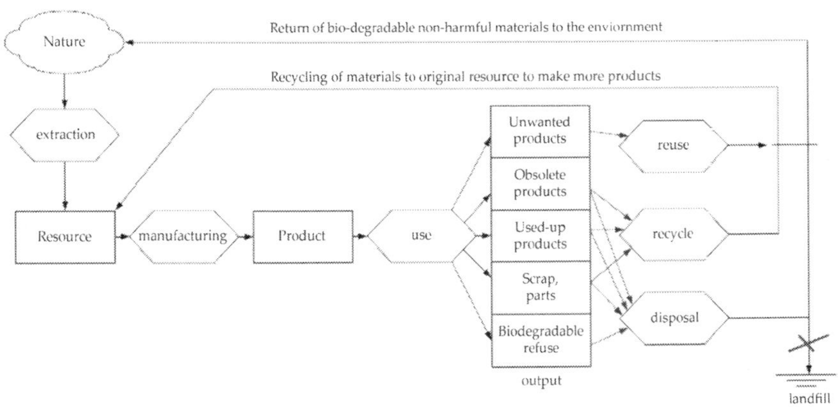

Figure 24: Closed-Loop Material Flow.

Source: Fedkin, M., (2017). Recycling: open-loop versus closed-loop thinking.

In this case, conversion of the used product back to raw material allows manufacturing of the same product repeatedly. Designing closed-loop systems, we need to remember some rules:

- The recycled materials should provide the same quality of the product without deterioration. For example, almost all recycled aluminum from soda cans is suitable to produce the same cans;
- Contaminants or toxins should not be accumulated in the

multiple recycling loop, as this can create an environmental risk the secondary product manufacturing;

- The recycled resource can also create the need in another manufacturing process for a different product or industry, which may require different type of recycling or couldn't be recycled at all (Fedkin, 2017).

The other part of closed-loop recycling concept is bio-degradable disposal. Everything that cannot be recycled or comes as a by-product in manufacturing process should be put back in the environment with no harm. While starting from the same extraction, manufacturing, and use stages, the closed-loop scheme become equally beneficial method for the manufacturing chain. With time flow, greater portion of materials should be designed for recycling and reuse (Fedkin, 2017).

Closed-loop approach is the goal of any sustainability scenario, but it would take radical changes and innovative thinking at the level of product and process design. Cyclical loop systems suggest that every product and all packaging should have a complete closed cycle designed for each component, i.e., pathways should be identified for each component to either be recycled indefinitely or to return to the natural ecosystem (Fedkin, 2017).

Dematerialization and Decarbonization

Economic development and industrial growth depend on growing levels of resource use, and result in environmental impacts from large scale resource extraction and emissions of waste. Economic growth, in addition to the desired, positive effects, entails negative side effects – environmental pollution, a global ecosystem thrown-off balance, widening standard-of-living differentials – which pose global challenges, thus forcing a revision of the prevailing patterns of thinking and acting. Many economists and sociologists point out that an aggressive consumerism is a path to quick depletion of the planet's unrenewable resources. Dematerialization and decarbonization are seen by Industrial Ecology as methods to design and realize sustainable manufacturing processes. Dematerialization represents the trust in the direction of decreasing intensity of material used by the societies as they develop that will help to produce more value with less resources and less waste. Decarbonization is an effort to reduce the amount of carbon-based fuels used in energy system and industrial manufacturing, and, thereby, decrease the amount of CO_2 emissions which cause the burning environmental issue of climate change.

Dematerialization. Defining Dematerialization, Robert Herman (1989) and his colleagues emphasized that we always should look at this process from environmental point of view. Using the less material could translate into smaller quantities of emissions and wastes generated at both the production and the consumption phases of the economic process. But less is not necessarily less from an environmental point of view. If smaller and lighter products are also inferior in quality, then more units would be produced, and the net result could be a greater amount of waste generated in both production and consumption. From an environmental viewpoint, therefore, dematerialization should be defined as "the change in the amount of waste generated per unit of industrial products" (Herman et.al., 1989).

Undoubtedly, many industrial products have become lighter and smaller with time. Cars, dwelling units, computers, calculators, and phones are but a few examples. Herman (1989) argued that production of smaller devices in some instances may result in lower-quality products and an increased consumer attitude to 'replace rather than repair'. In these instances, the number of units produced may have increased. Although dematerialization may be the case on a per-unit basis, the increasing number of units produced can cause an overall trend toward materialization with time. As an example, the apparent consumption of shoes, which seem increasingly difficult to repair, has risen markedly in the United States since the 1970s (Herman et.al., 1989).

In contrast, improvements in quality generally result in dematerialization. The well-known example of this is the case for automobile tires. Statistics tells us that total tire production in the United States has risen over time, following from general increases in both the number of registered vehicles and the total miles of travel. However, the number of tires per million vehicle miles of travel has declined attributed to improved tire quality, which results directly in a decrease in the quantity of solid waste due to discarded tires (Herman et.al., 1989).

Dematerialization of some production process affects, and is influenced by, several factors besides product quality. According to Herman (1989), these include ease of manufacturing, production cost, size, and complexity of the product, whether the product is to be repaired or replaced, and the amount of waste to be generated and processed. These factors influence one another as well, as it is shown at the Figure 25.

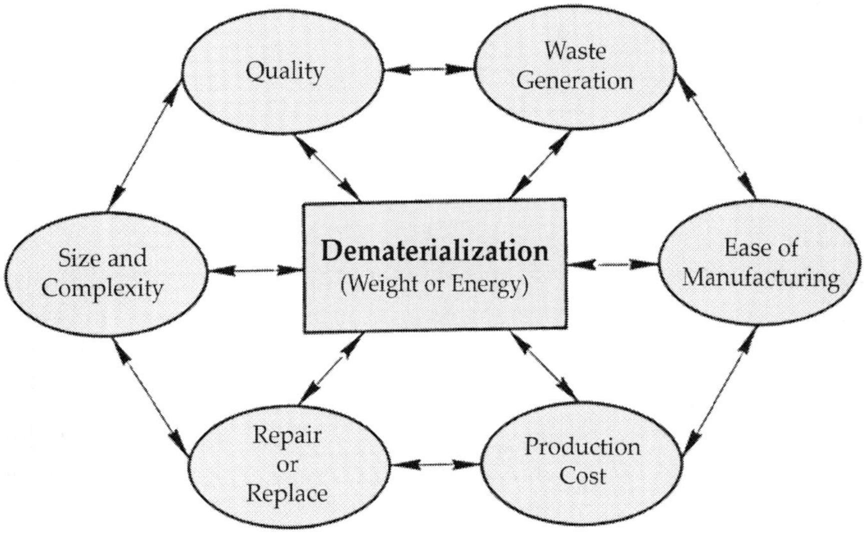

Figure 25: Factors Affecting, and Affected by, the Dematerialization Process.

Source: Robert Herman, Siamak A. Ardekani, and Jesse H. Ausubel Dematerialization. Program for the Human Environment, The Rockefeller University http://phe.rockefeller.edu/dematerialization

Economic and population growth, of course, also strongly interact with many of the factors. For example, the ease of manufacture of a particular product in smaller and lighter units may result in lower production cost and cheaper products of lower quality, which will be replaced rather than repaired on breaking down. Although a smaller amount of waste will be generated on a per-unit basis, more units will be produced and disposed of, and there may be an overall increase in waste generation at both the production and the consumption ends (Herman et.al., 1989).

In 1989 Dr. Herman predicted that "spatial dispersion of population is a potential materializer. Migration from urban to suburban areas, often driven by affluence, requires more roads, more single-unit dwellings, and more automobiles with a consequent significant expansion in the use of materials. The movement from large, extended families sharing one dwelling to smaller, nuclear families may be regarded as a materializer if every household unit occupies a separate dwelling. Factors such as photocopying, photography, advertising, poor quality, high cost of repair, and wealth generally force materialization. Technological innovation, especially product innovation, may also tend to force materialization, at least in the short run. National security and war, styles and fashions, and fads may also function as materializers by accelerating production and consumption. Demand for health and fitness, local mobility, and travel may spur materialization in other ways" (Herman et.al., 1989). Today all these predictions have been realized.

Nowadays we can observe socioeconomic factors influencing society's demand for various activities, composition of the work force, and income levels. Developing countries evolves toward industrialization, increasing their demand for materials, whereas developed countries are on the way of their transition from an industrial to a service society, that might bring about a decline in the use of materials. Within a given culture, we can't predict for sure to what amount of materials use and waste generation increasing functions of income will lead (Herman et.al., 1989).

In countries with high level of economic and industrial development, some theories state that a combination of factors could lead to the absolute dematerialization and decarbonization of the economy. These factors are generally described as:

- structural shifts in production and consumption patterns (often known as the transition to a 'service economy', although this transition is disputed (Kander, 2005);
- high technical capacity (leading to more efficient use of resources);
- surplus resources to devote to environmental protection (Stern, 2004);
- economic preferences for outsourcing labor-intensive (and possibly pollution-intensive or primary commodity-intensive) activities (Muradian and Martinez-Alier, 2001).

Working in a complex, these factors could be expected to lead to an 'Environmental Kuznets Curve' (EKC), where environmental impacts first grow, then decrease with income, in the shape of an inverted-U. The EKC suggests that beyond a certain level of per-capita income, adverse environmental trends begin to reverse themselves. Past studies have approved the EKC behavior for certain categories of pollutants, which tend to have in common that their impacts are local and immediate, while their abatement is technically straightforward and low cost (Stern, 2004; Smith and Ezzati, 2005). A stylized representation of the EKC takes the form of an inverted U, with income increasing horizontally and pollution rising vertically. It shows us that developing countries, in their initial growth tendency, become still more polluted, but with economic and technological progress they begin reversing that trend as they recognize the benefits of a cleaner environment and green economy.

The idea of dematerialization in modern society is to minimize total material and energy consumption while ensuring the services which are at least equivalent. Technical progress makes it possible to obtain more services with less resources. More generally, one of the best ways to dematerialize the economy is to optimize usage: functionally and from the point of overconsumption.

In other words, producer should try to sell the usage rather than the object (this is referred to as the functionality economy); for example, it is in the best interest of a photocopier manufacturer selling a 'photocopying' service, instead of the actual machine, to ensure that their photocopier requires as little material as possible, has the longest possible functional shelf life, is easy to recycle etc.

Dematerialization, according to Walter Stahel is based on the strategy of product life extension (Stahel, 2013):

- Reuse;
- Repair;
- Remanufacture;
- Recycle (the last option!).

Although no recycling process is 100 % efficient, recycling is a promising means of dematerialization. Another question to be raised about the economics of consumption and waste disposal is what the 'true' cost of consumption and processing of the generated waste is to society. What is the true cost of burning fossil fuel for transportation while we are takin to account the finiteness of resources and consequent long-term damage to the environment? Should these non-renewable fossil fuels be made available at much higher cost so that profits may be reinvested for development of the capital and the knowledge to permit the use of renewable resources and the development of technological substitutes? What is the true cost of municipal and industrial wastes disposal? Would a higher cost for garbage disposal effectively encourage recycling, sorting recyclable materials at the generation source, and dematerialization? Or it could be resulted in illegal dumping? Can society truly afford to continue functioning in its present 'throwaway' model with products such as food, clothing, diapers, shoes, watches, radios, light bulbs, cameras, calculators, pens and pencils, razors, etc.? (Southerstone et al., 2004; Stahel, 2013).

Industrial Ecology has increasingly questioned he model of unsustainable consumption in favor of a sustainable model, where people draw on consumer goods and services with restraint and a sense of responsibility for future generations. Calls for embracing sustainable development, to be accompanied by a moderate and responsible consumption, have therefore multiplied, requesting conscious consumers of the new type to demonstrate greater sensitivity and responsibility in respect of the environment, and buy only such products that do not involve energy-intensive production methods and do not lead to resource depletion, environmental pollution, etc. (Southerstone et al., 2004). Coming as an essential complement to sustainable development is sustainable consumption, which could be described through the following characteristics:

- a balanced relationship between consumption and savings;
- stable living standards and consumers' material situation;
- widespread environmental awareness;
- a limited number of people who consume while not working (harmony between the individual's roles as producer and consumer);
- prices of eco-friendly goods and services taken into account in consumers' budgets;
- an appropriate pattern of material consumption on the one hand and, on the other, direct consumption of eco-friendly products and non-material components of welfare;
- emphasis on such forms of consumption, which involve low environmental and social costs;

- following, in practice, the principle of inter- and intra-generational justice; growing welfare of some consumer groups must not restrict welfare-improvement opportunities for other groups and next generations (Kiełczewski, 2008).

Promotion of sustainable consumption is of paramount importance for the environment: assuming that one day the consumption levels of the highly-developed world would be reached by all countries, this would require ten times as much of natural resources as the planet is capable of supplying. As pointed out by the environmentalists, the oft-embraced imperative of growth at any price, stimulated by a continuous arousal of consumer aspirations, leads to the depletion of unrenewable natural resources and upsets a delicate balance of the global ecosystem. It is, therefore, demanded that consumption should shift towards greater environment-friendliness, as reflected in the following behaviour (Janoś-Kresło and Mróz, 2006):

- using consumer goods in a frugal and rational manner;
- purchasing less of such goods whose consumption involves the use of unrenewable raw materials and creates hazardous or toxic post-consumption waste;
- purchasing and using such consumer goods, which do the least damage to the environment and human health;
- consuming eco-friendly goods (especially in food production), without chemical additives, preservatives, etc.

The concept of sustainable consumption involves a strategy to influence consumer demand and behavior towards such a strategy of natural resources use and economic capacity utilization, which ensures that needs are met and quality of life raised for all consumers, while at the same time reproducing the nature's capital for the benefit of future generations. The consumption strategy meeting these criteria poses no major threats to the environment and discourages an intense exploitation of natural resources is referred to in literature as green consumption, ecology-friendly consumption, or eco-consumption (Kiełczewski, 2008).

While accounting for just a small fraction of global consumption, it nevertheless offers a promising sign of the contemporary consumer's increasing environmental awareness and will no doubt be gaining in importance. This expectation is confirmed by the growing demand for clean products, which pose as little burden on the environment as possible, by requiring low raw-material and energy inputs and not involving large amount of troublesome and costly waste (Southerstone et al., 2004). Such products are safe for human health and are made from renewable and reusable resources. The market for organic, eco-labelled products has been expanding dynamically, especially in sectors such as household tools and stuff (refrigerators, freezers, washing machines, dish washers), paints and dyestuffs, detergents, household chemicals, which are important for people's health and family budgets. These trends should be accompanied with other methods, such as prolonging products' life-cycles and forcing producers to give more consideration to the durability of products on offer (Kiełczewski, 2008).

***Decarbonization*:** During all the industrial development history, carbon in the form of fossil hydrocarbon fuels – coal, petroleum, gas – represents the core element, the vital substance irrigating all economies developing in the world. However, this fossil carbon is the source of countless problems: intensification of the greenhouse effect and climate change, smog, oil spills, acid rain, etc. Therefore, the harmfulness of hydrocarbon consumption must be reduced and the transition to an energy mix involving less fossil carbon must be encouraged, extending the usage of renewable energy and energy saving technologies.

Decarbonization of energy systems relies on four pillars (Stahel, 2013):

- ambitious energy efficiency;
- low carbon electricity;
- electrification and fuel switching;
- reducing non-energy emissions in industry and agriculture.

The Deep Decarbonization Pathways Project (DDPP, 2015) is a collaborative global research initiative seeking to understand how individual countries can transition, on a technological, socio-economic and policy 'pathway', to a low-carbon economy consistent with the internationally agreed goal of limiting anthropogenic warming to less than 2 degrees Celsius (2°C). Achieving this goal requires that the world cut global net emissions of greenhouse gases so that they approach zero between 2050 and 2075. According to the Intergovernmental Panel on Climate Change findings, "a better-than-even chance of remaining below a 2°C temperature rise, global annual emissions will need to be reduced 42%–57% by 2050 (relative to 2010), and 73%–107% by 2100" (IPCC, 2015).

This 'deep decarbonization' transition will entail, as it is shown at the Figure 26, more than any other factor, the profound transformation of energy systems through steeply reducing carbon intensity in all sectors of the economy (DDPP, 2015).

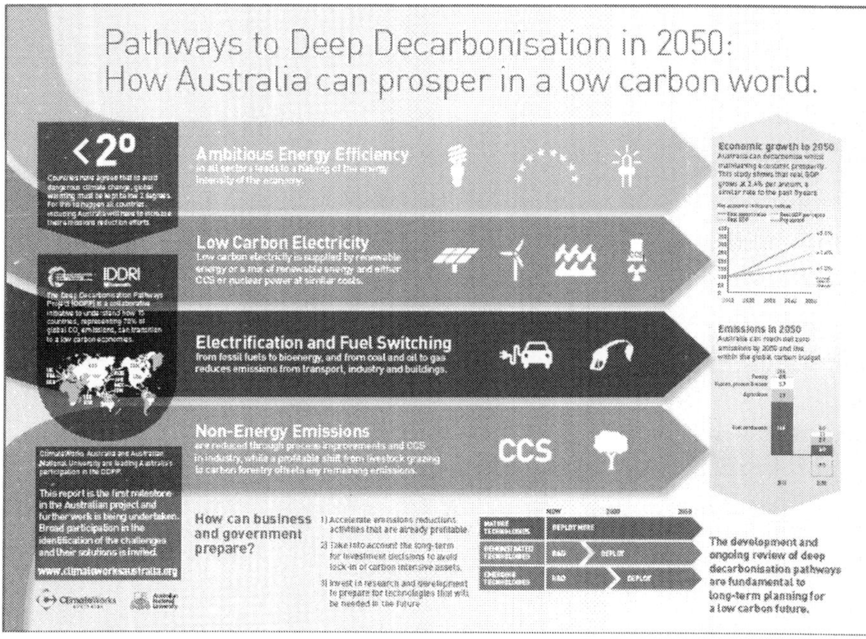

Figure 26: Pathways to Deep Decarbonization.

Source: Deep Decarbonization Pathways Project (DDPP), (2015). Available at http://deepdecarbonization.org/2015/07/pathways-to-decarbonization-towards-a-2050-low-carbon-australia/

Dematerialization and decarbonization, overlapping with the EKC phenomenon, refer to advanced societies' decreasing dependence on fossil fuels and energy-intensive manufacturing sectors as their economies become more high-tech and service oriented. Scientists exploring dematerialization and decarbonization begin with dissecting past trends and then ponder the implications for the future. After analysis was made, we can surely understand that, even though the empirical base is sturdier and extends over a much longer period, no inexorable future development can be divined from these concepts, especially decarbonization (Darmstadter, 2010).

Environmental Impact Assessment

Environmental Impact Assessment (EIA) can be defined as "the systematic identification and evaluation of the potential impacts (effects) of proposed projects, plans, programs, or legislative actions relative to the physical, chemical, biological, cultural and socio-economic components of the total environment" (Council of Environmental Quality, 1993). The primary purpose of the EIA process is "to encourage the consideration of the environmental issues in planning and decision making and to ultimately arrive at actions which are more environmentally compatible" (Canter, 1996).

According to Fulekar (2015), "the purpose of Environmental Impact Assessment is to enhance projects by helping present, minimize, mitigate or compensate for any adverse environmental and social impacts. Development institutions and many governments have introduced EIA requirements and regulations into their development activities".

For Industrial Ecology, environmental impact assessment can be considered as a mechanism which maximizes the efficient use of natural and human resources; reduce costs and time taken to reach a decision by ensuring that subjectivity and duplication of effort are minimized; as well as identifying and attempting to evaluate the primary and secondary consequences which might require the introduction of expensive pollution control equipment or compensation and other costs later (Mondal, 2016).

The potential scope of an inclusive EIA system is considerable and could contain the appraisal of policies, plans, programs, and specific projects. EIA, as it has developed in many countries, is based on several procedures and stages (Mondal, 2016):

- Identification of projects requiring EIA, sometimes known as screening;
- Identification of the key issues to be addressed in an EIA, called scoping;
- Impact assessment and evaluation;
- Impact mitigation and monitoring;

- Review of the completed Environmental Impact Statement and;
- Public participation.

Charles Eccleston (2011) defines two roles of environmental impact analysis:

- Legal role – Is quite straight forward to ensure that development projects such as a housing estate, a road/bridge or some such construction project has a minimal impact on the environment in its entire 'lifecycle' – i.e. during design, construction, use, maintenance, and demolition. Many countries now have laws stipulating that unless an EIA study is carried out particularly for large infrastructure projects, permission for construction will not be granted by the local authority;
- Educational role – is equally important to educate everyone involved both professionals and users. We need to look at all our daily actions as eventually and cumulatively affecting the environment. This includes our daily choices, where a delicate balance between financial and environmental considerations needs to be made automatically without thinking (Eccleston, 2011).

The result of an environmental impact analysis is accumulated in a document known as an Environmental Impact Statement which represents all the positive and negative effects of a particular project on the environment. This report is just one component of the information required to aid decision makers in managing the project (Fulekar, 2015). Environmental Impact Assessment Statement covers the following five major areas:

- The environmental effects of the proposed action;
- Any adverse environmental effects that cannot be avoided. And question if, in such case, the proposal should be implemented;
- Alterations to the proposed action;
- The relationship between the local short-term uses of the human environment and the support and enhancement of the long-term efficiency.

- Any irreversible and irretrievable commitment of resources that would be elaborate in the proposed action should be implemented (Fulekar, 2015).

Holder (2004) described the five main components under investigation during environmental impact analysis:

1) Air:
    - Air quality;
    - Wind direction / speed;
    - Precipitation / humidity;
    - Temperature;
    - Noise.
2) Land:
    - Land capability;
    - Soil resources / structure;
    - Mineral resources;
    - Tectonic activity;
    - Unique features.
3) Water
    - Surface water;
    - Ground water regime;
    - Hydrologic balance;
    - Drainage / channel pattern;
    - Flooding;
    - Sedimentation.
4) Flora and fauna:
    - Environmentally sensitive area;
    - Wetlands, marshes, wild lands, grasslands etc.;
    - Species inventory;
    - Productivity;
    - Biogeochemical / nutrient cycle.
5) Human:
    - Social infrastructure / institutions;

- Cultural characteristics;
- Physiological and Psychological well-being;
- Economic resources (Holder, 2004).

US Environmental Protection Agency developed a set of Environmental Impact Assessment Criteria (Table 2):

Table 2: Environmental Impact Assessment Criteria

Ranking	Classification	Impact on the environment
1	Minor	***No contamination, localized effects*** • Minor effect on air quality as evidenced by dust or odour complaint(s) • ELV breaches • An emission which does not comply with the requirement of the licence/COA (A pattern of repeated minor incidents should be taken into account when considering the level of response)
2	Limited	***Simple contamination, localized effects of short duration*** • Local limited impact to water, land and air • Notification to and short term closure of potable water extractors required
3	Serious	***Simple contamination, widespread effects of extended duration*** • Significant effects on water quality • Major damage to an ecosystem (e.g. significant impact on fish population) • Longer term closure of potable water extractors • Significant reduction in amenity value • Significant Damage to agriculture or commerce • Significant Impact on man

4	Very Serious	***Heavy contamination, localized effects of extended duration***
5	Catastrophic	***Very heavy contamination, widespread effects of extended duration***

Source: EPA, https://www.epa.ie/pubs/advice/licensee/Guidance%20to%20licensees.pdf

Environmental Impact Assessment can be classified based on the objective and topic of development (Eccleston, 2011):

- climate impact assessment;
- demographic impact assessment;
- development impact assessment;
- ecological impact assessment;
- economic and fiscal impact assessment;
- health impact assessment;
- risk assessment;
- social impact assessment;
- strategic impact assessment;
- technology assessment.

In addition to this list, Environmental Impact Assessmant is also categorized based on systematic analysis of ecological characteristics, geographical region, carrying capacity limitations, and sectoral planning. They are:

***Strategic EIA*:** Strategic EIA refers to regular analysis of the environmental effects of development policies, plans, programs and other proposed tactical actions. This process outspreads the aims and principles of EIA upstream in the decision- making process, beyond the project level and when major alternatives are still open. Strategic EIA represents a proactive approach to integrate environmental considerations into the higher level of decision-making (Mondal, 2016).

***Regional EIA*:** EIA in the context of regional planning integrates environmental concerns into development planning for a geographic region, normally at the sub- country level. Such an approach is referred to as the 'economic-cum-environmental' development design. This approach facilitates adequate incorporation of economic development with management of renewable natural resources within the carrying capacity limitation to achieve sustainable development. It fulfills the need for macro-level environmental integration, which the project-oriented EIA is unable to address effectively. Regional EIA addresses the environmental impacts of regional development plans and thus, the context for project-level EIA of the subsequent projects, within the region. In addition, if environmental effects are considered at regional level, then cumulative environmental effects of all the projects within the region can be accounted (Mondal, 2016).

***Sectoral EIA*:** Instead of project-level-EIA, an EIA should take place in the context of regional and sectoral level planning. Once sectoral level development plans have the integrated sectoral environmental issues addressed, the scope of project-level EIA will be specified. Sectoral EIA will help to address specific environmental problems that may be encountered in developing and implementing sectoral industrial projects (Mondal, 2016).

***Project Level EIA*:** Project level EIA refers to the designing activity in isolation and the impacts that it exerts on the receiving environment. Thus, it may not effectively integrate the cumulative effects of the development in a region (Mondal, 2016).

***Life Cycle Assessment*:** A broader approach to deal with environmental impacts in manufacturing is called life cycle assessment. This approach recognizes that environmental concerns enter into every stage of the process with respect to the manufacturing, of the products and thus examines environmental impacts of the product at all phases of the product life cycle (Mondal, 2016).

From the above discussion, it is clear that EIA shall be integrated at all the levels i.e. strategic, regional, sectoral and the project level. Whereas, the strategic EIA is a structural change, the regional EIA refers to substantial information processing and drawing complex inferences. The project-level EIA is relatively simple and reaches to meaningful conclusions. As we progress and the resource planning concepts emerge in our decision- making process, the integration of overall regional issues will become part of the impact assessment studies (Mondal, 2016).

The fundamental approach for EIA, according to Fulekar (2015), is to conduct study with the fundamental issue being to determine whether the predicted impacts of the action would have a significant effect on the quality of the human environment. Impacts resulting from industrial activity can be placed in one or more of the following categories:

- Beneficial or detrimental.
- Naturally reversible or irreversible.
- Responsible via management – practices or irreparable.
- Short term or long ter.
- Temporary or continuous.
- Occurring during the construction phase or the operational phase.
- Local, regional, national or global.
- Accidental or planned.
- Direct (primary) or indirect (secondary).
- Cumulative or single (Fulekar, 2015).

'Environment' in EIA context mainly focuses physical, chemical, biological, geological, social, economic, and aesthetic dimensions along with their complex interactions, which affects individuals, communities and ultimately determines their forms, character, relationship, and survival (Fulekar, 2015).

According to International Organization for Impact Assessment, indicators for EIA can be classified in to environmental performance indicators and environmental condition indicators. The environmental performance indicators can be further divided into two categories namely, operational performance indicators and management performance indicators. The operational performance indicators are related to the process and other operational activities of the organization, and would typically address the issue of raw material consumption, energy consumption, water consumption in the organization, the quantities of waste water generated, other solid wastes generated, emission from the organization (IAIA, 1999).

Dr. M.H. Fulekar outlines the major five steps of environmental impact assessment:

- Screening
- Scoping and Development of Terms of reference

- Preparing the Environmental Assessment Report
- EA review and Project Appraisal
- Project Implementation (Fulekar, 2015)

Let us analyze every of these stages.

Stage 1. Screening: The team expert assesses the project to determine the nature and magnitude of the proposed projects potential environmental and social impacts. This falls into three categories (Fulekar, 2015):

- *Category A.* Full EIA is required Category A project are those expected to have adverse impacts that may be sensitive, irreversible and diverse, with attributes such as direct Pollutant Discharges large though to cause degradation of air, water, or soil, physical disturbance of site or surroundings and natural resources;
- *Category B.* Although a full EIA is not required, some environmental analysis is necessary. This type of projects has the impacts which are less significant, not as sensitive, numerous, major, or diverse. The Category B projects entail rehabilitation, maintenance or upgradation rather than new constitution;
- *Category C.* No EIA or other environmental analysis is required. This category projects fall on education, family planning, health, and human resource development. This category projects entail negligible or minimal direct disturbance to physical setting.

Stage 2. Scoping and Development of Terms of Reference: At this stage, it is essential to identify mere precisely the likely environmental impacts. The information about the project and its likely environmental effects is disseminated to local affected communities and Non-Governmental organizations (Fulekar, 2015).

Stage 3. Preparation of Environmental Impact Assessment Report:

- *Category A – Project* – a full scale Environmental Impact Assessment is normally undertaken, resulting in EIA report;
- *Category B – Project* requires limited EIA, differ from case to case;

- *Category C – Project* – no EIA is required (Fulekar, 2015).

The main components of a full EIA reports include:

- Executive Summary;
- Policy, legal and administrative full work;
- Project description;
- Baseline data;
- Impact Assessment;
- Analysis of alternatives;
- Initiation or Management Plan;
- Environmental monitoring Plan;
- Public consultation (Fulekar, 2015).

Stage 4. Environmental Assessment Review and Project Appraisal. Once the draft EIA report is complete, the project team is authorized to proceed appraisal of the project. On the appraisal mission, the project authority review the EIA's procedural and substantive elements with the proper authority, resolve any outstanding issues, assess the adequacy of the institutions for environmental management in the light of the EIA's findings, ensure that the mitigation plan is adequately budgeted (Fulekar, 2015).

Stage 5. Project Implementation. The project authority is responsible for implementation of the project according to the agreements derived from the EIA processes. The environmental experts supervise the implementation of the project (Fulekar, 2015).

Life Cycle Assessment

As a concept tool of Industrial Ecology, Life Cycle Assessment (LCA) is a method of evaluating the environmental consequences of a product or process 'from cradle to grave' (Vigon et.al., 1993). The Society for Environmental Toxicology & Chemistry (SETAC) defines LCA as "a process used to evaluate the environmental burdens associated with a product, process, or activity" (SETAC, 1993).

US Environmental Protection Agency defines Life-Cycle assessment (LCA) as a tool that can be used to evaluate the potential environmental impacts of a product, material, process, or activity. An LCA is a comprehensive method for assessing a range of environmental impacts across the full life cycle of a product system, from materials acquisition to manufacturing, use, and final disposition. LCA study results help to promote the responsible design and redesign of products and processes, leading to reduced overall environmental impacts and the reduced use and release of more toxic materials. LCA studies identify key materials and processes within the products' life cycles that are likely to pose the greatest impacts, including occupational and public toxicity impacts. These assessments allow businesses to make product improvements through environmentally sound process, material, and design choices. In the United States, SETAC, the U.S. EPA and consulting firms are active in developing LCAs (EPA, 2017).

According to ISO 14040, Life Cycle Assessment is "a systematic set of procedures for compiling and examining the inputs and outputs of materials and energy and the associated environmental impacts directly attributable to the functioning of a product or service system throughout its life cycle". LCA evaluates the potential environmental aspects and potential aspects associated with a product (or service), by:

- compiling an inventory of relevant inputs and outputs,
- evaluating the potential environmental impacts associated with those inputs and outputs,
- interpreting the results of the inventory and impact phases in relation to the objectives of the study (ISO 14040).
- Life-cycle assessment has emerged as a valuable decision-support tool for both policy makers and industry. Three forces were driving this evolution (Ang et.al., 2014):
- First – government regulations that are moving in the direction of 'eco-accountability' – the concept that a manufacturer is responsible not only for direct manufacturing impacts, but also for impacts associated with product inputs, use, transport, and disposal;
- Second – business participation in voluntary initiatives which contain LCA and product stewardship components (i.e., ISO 14000, Chemical Manufacturer Association's Responsible Care Program), which seek to foster constant improvement through better environmental management systems;

- Third – environmental 'preferability' that has emerged as a criterion in government procurement and consumer markets guidelines (Ang et.al., 2014).

Together these developments have placed Life Cycle Assessment in a central role as an Industrial ecology tool for identifying impacts of products and the materials from which they are made and designing sustainable manufacturing processes. The main goal, or 'ideal' of Life Cycle Assessment is to cycle the production process, so no raw materials are used and no wastes are disposed into the biosphere.

Life cycle thinking is a very important principle in industrial ecology. It implies that all environmental influences caused by a manufacturing process, product, or project during its life cycle are considered (Figure 27). In this context life cycle assessment analyses (Ang et.al., 2014):

- Raw material extraction;
- Material processing;
- Manufacture;
- Use;
- Maintenance;
- Disposal;
- Transportation;
- Reuse;
- Remanufacture;
- Recycle.

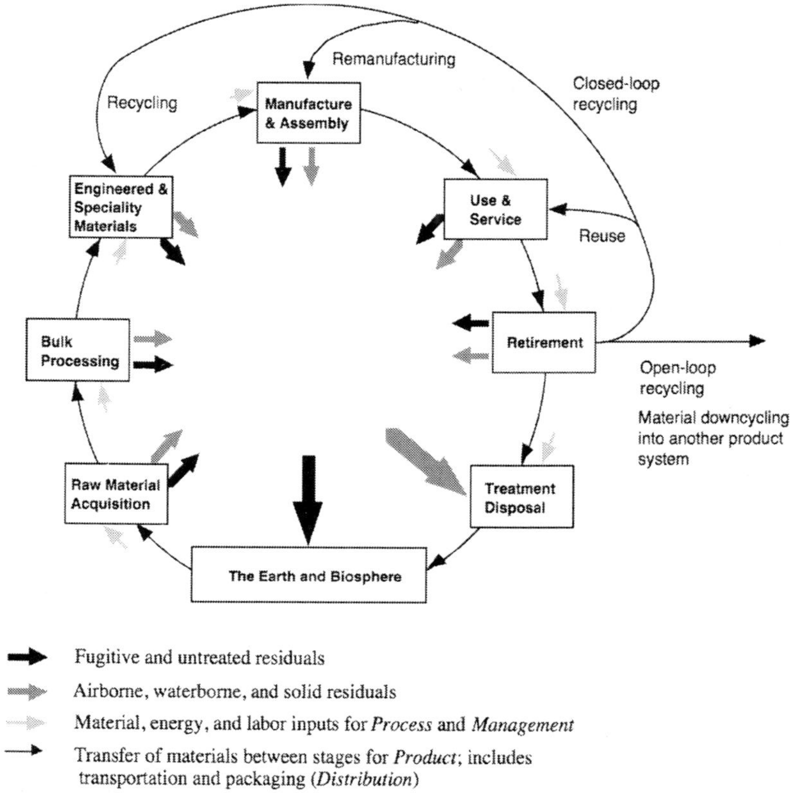

Figure 27: Life Cycle Assessment Process.

Source: Gregory A. Keoleian and Dan Menerey, Life Cycle Design Guidance Manual (Cincinnati: U.S. EPA Risk Reduction Engineering Lab, 1993), 14.

Life Cycle Assessment is based on four linked components, according to the ISO 14040 Standard (Figure 28):

- Goal definition and scoping;
- Life-cycle inventory;
- Impact analysis;
- Improvement analysis.

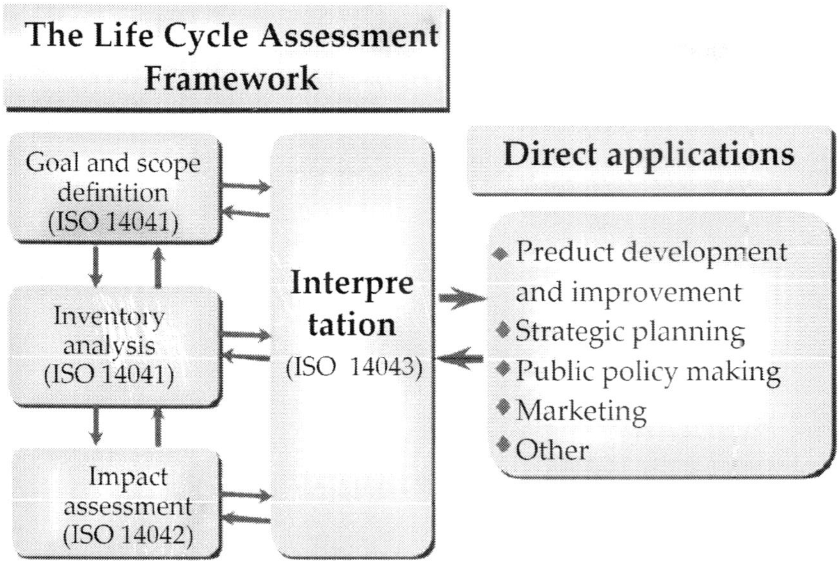

Figure 28: The Framework of LCA with Four Interrelated Phases.

Source: Dašić, P. et.al., (2009) Software systems and management technology in environment Available at: https://www.researchgate.net/publication/269811337_Software_systems_and_management_technology_in_environment.

The first step, **Goal and Scope definition**, is the explicit specification of goal and scope of the LCA (Bauer and Treyer, 2015). In this section, the most important choices of the study are described in detail, i.e. methodological choices, assumptions and limitations, particularly with regards to the following issues:

- specification of the product system to be investigated;
- system boundaries;
- functional unit(s);
- reference flow(s);
- assumptions and limitations;

- procedures for allocation or system expansion;
- impact categories chosen (Bauer and Treyer, 2015).

The goal and scope definition is a guide that ensures the LCA is performed consistently. SimaPro software proposes such a scheme for defining goal and scope of assessment (Figure 29).

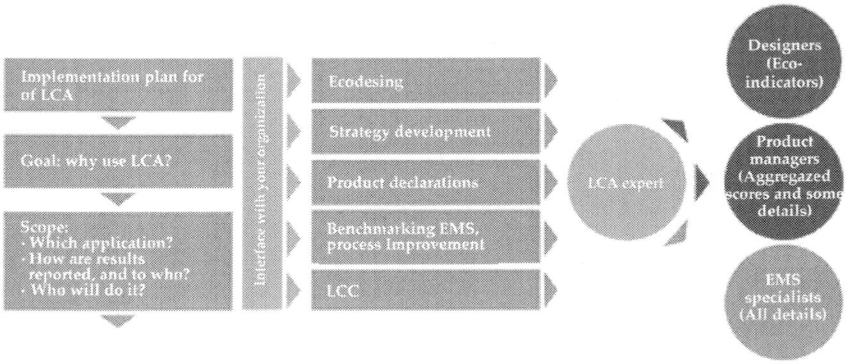

Figure 29: The Choice of Goal and Scope Definition in SimaPro.

Source: Goedkoop et.al., (2016). Introduction to LCA with SimaPro. Language. Available at: https://www.pre-sustainability.com/download/SimaPro8IntroductionToLCA.pdf

The second phase, **Inventory Analysis (LCI),** comprises the creation of inventories of flows from and to nature and the technosphere (i.e., the 'man-made world') for a product system. Inventory flows include inputs of water, land, energy, raw and processed materials, and releases to air, land, and water (Figure 30). The product system (i.e., all processes being part of a product life cycle) can be divided into 'foreground and background' system. The distinction is often ambiguous, but in general, the forefront system represents those processes, which are directly part of a value chain of the product or services in focus of the LCA, while the background system represents all up- and downstream processes related to the foreground system (Bauer and Treyer, 2015).

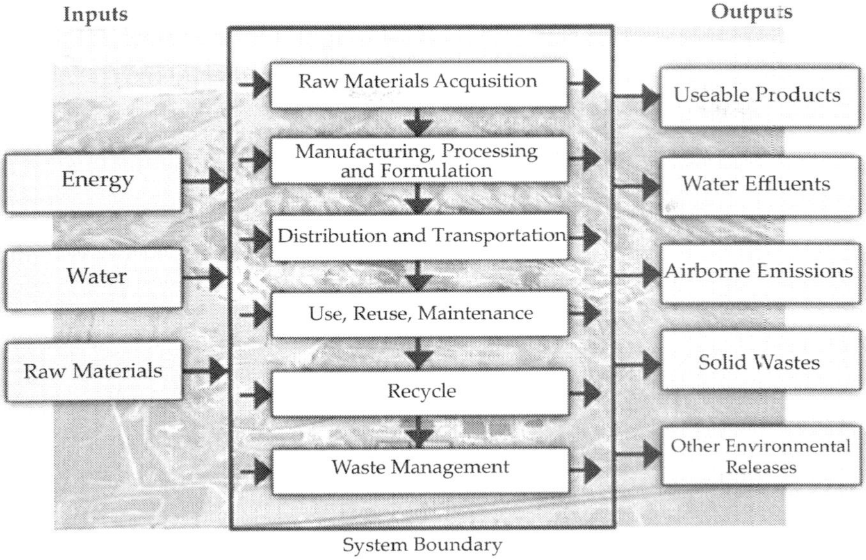

Figure 30: Inputs and Outputs Over a Product's Life Cycle.

Source: Berkeley Lab. (2008). Life-Cycle Analysis. Available at: http://newscenter.lbl.gov/2008/04/18/life-cycle-analysis/

The background system usually comprises thousands of processes and therefore, generic background LCI databases such as the eco-invent database are used. Based on the inputs and outputs of all individual processes, cumulative LCI data can be calculated representing the life-cycle environmental burdens per functional unit of the reference product (Bauer and Treyer, 2015).

The inventory process seems simple enough in principle. In practice, however, it is subject to a number of practical issues, such as geographical variations, data quality, and choice of technology (Goedkoop et.al., 2016).

The third phase of LCA, **Impact Assessment (LCIA),** aims at evaluating the importance of potential environmental impacts based on the cumulative LCI results. Traditional LCIA consists of the following mandatory elements (Bauer and Treyer, 2015):

- selection of impact categories, category indicators, and characterization models;
- the classification stage, where the inventory flows are sorted

and assigned to specific impact categories;

- quantification of impacts, where the LCI flows are characterized, using one of many possible LCIA methodologies, into common equivalence units.

LCA manual with SimaPro software admits that for each substance, a schematic cause response pathway needs to be developed that describes the environmental mechanism of the substance emitted. Along this environmental mechanism, an impact category indicator result can be chosen either at the midpoint or endpoint level (Goedkoop et.al., 2016).

Midpoint impact category, or problem-oriented approach, translates impacts into environmental themes such as climate change, acidification, human toxicity, etc. Endpoint impact category, also known as the damage-oriented approach, translates environmental impacts into issues of concern such as human health, natural environment, and natural resources. Endpoint results have a higher level of uncertainty compared to midpoint results but are easier to understand by decision makers (Goedkoop et.al., 2016).

Figure 31 shows us an example of characterization process in SimaPro software. The first three substances all contribute to climate change, CFCs only contribute to ozone layer depletion, and NO2 contributes to eutrophication. Having three substances contributing to one impact category, does not mean that we can now say we have 1,101 grams of climate changing substances, as these three substances all contribute differently to the problem of climate change. To determine this, the IPCC equivalency factors were used. Similarly, for other impact categories different characterization factors have been defined. Each LCI result is multiplied with a characterization factor before they can be added. The characterization factor is determined by the impact assessment method (Goedkoop et.al., 2016).

Elementary flow	Climate change		Ozone layer depletion		Eutrophication	
	CF	Result	CF	Result	CF	Result
1 kg CO_2 (carbon dioxide)	1	1				
10 gram CH_4 (methane)	25	0.25				
1 gram CFC142b	2310	2.31	0.07	0.00007		
5 gram NO_2					0.56	0.0028
Impact Category Indicator Result		3.56		0.00007		0.0028
Unit of the result	kg CO_2 equivalent		kg CFC11 equivalent		kg P equivalent	

Figure 31: The Result of Characterization in SimaPro.

Source: Goedkoop et.al., (2016). Introduction to LCA with SimaPro. Language. Available at: https://www.pre-sustainability.com/download/SimaPro8IntroductionToLCA.pdf

The table at Figure 31 illustrates several issues of the characterization phase:

- One emission can contribute to several impact categories. Although the ISO standard does mention the opportunity to split the elementary flow over two or more impact categories, this is generally not done. One substance can cause several things at the same time;
- In this (and many other so-called midpoint methods) the units are defined as a reference to a substance. This results in relatively nonconcrete units like CO_2 equivalents;
- It is very important to note that the units of the results are different, thus impact categories cannot be compared to each other and the overall scale of impacts cannot be determined. Due to the rather arbitrary units, the characterization results are not always easy to interpret (Goedkoop et.al., 2016).

The fourth phase, **Interpretation**, aims to formulate a set of conclusions and recommendations based on the LCA results. Interpretation, according to ISO 14044, describes several checks you need to make to ensure the conclusions are adequately supported by the data and procedures used in the study. The results are reported in the most informative way possible and the need and opportunities to reduce the impact of the product(s) or service(s) on the environment are systematically evaluated (Goedkoop et.al., 2016). The following checks are recommended:

- Uncertainty;
- Sensitivity analysis;
- Contribution analysis (Goedkoop et.al., 2016).

These four phases are often interdependent, i.e. LCA is carried out in an iterative way. The scientific community distinguishes between 'attributional' and 'consequential' LCA (Vigon et.al., 1993; Bauer and Treyer, 2015). Attributional LCA (aLCA) pursues to attribute the burdens associated with the manufacturing and use of a product, or with a specific service or process, at a certain point in time (typically the recent past). LCA represents the average mix of market suppliers or supply technologies. Consequential LCA (cLCA) seeks to identify the environmental consequences of a decision or a proposed change in the system under study (oriented to the future). Therefore, market information and economic implications of a decision have to be taken into account (Bauer and Treyer, 2015).

The example overview of the life cycle of a fax machine is depictured at Figure 32.

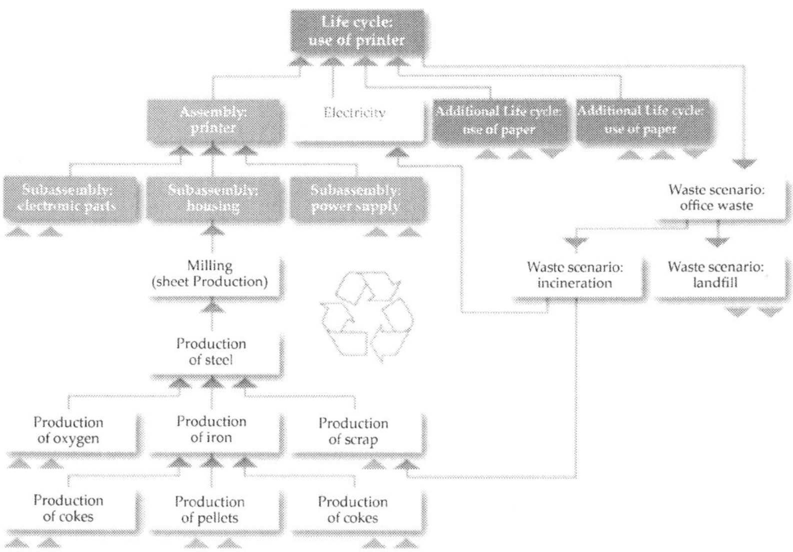

Figure 32: Schematic Overview of the Life Cycle of a Fax Machine.

Source: Goedkoop et.al., (2016). Introduction to LCA with SimaPro. Language. Available at: https://www.pre-sustainability.com/download/SimaPro8IntroductionToLCA.pdf

At the Figure 32, the grey boxes are product stages, the white boxes are processes. As we can see, a life cycle can link up to:

- One assembly (which may have subassemblies).
- One or more use processes, in this case electricity.
- One or more life cycles of auxiliary products, such as paper and cartridges. These auxiliary life cycles are defined just as any other life cycle. They also have an assembly and an end-of-life stage. This allows to model a different end-of-life scenario for paper, cartridges and fax machines.
- A waste or disposal scenario (in this case a waste scenario is assumed).
- Additional life cycles (Goedkoop et.al., 2016).

Christian Bauer and colleagues developed the product system for the LCA of electric vehicles battery which is shown ar the Figure 33 (functional unit: 1 vehicle-kilometer).

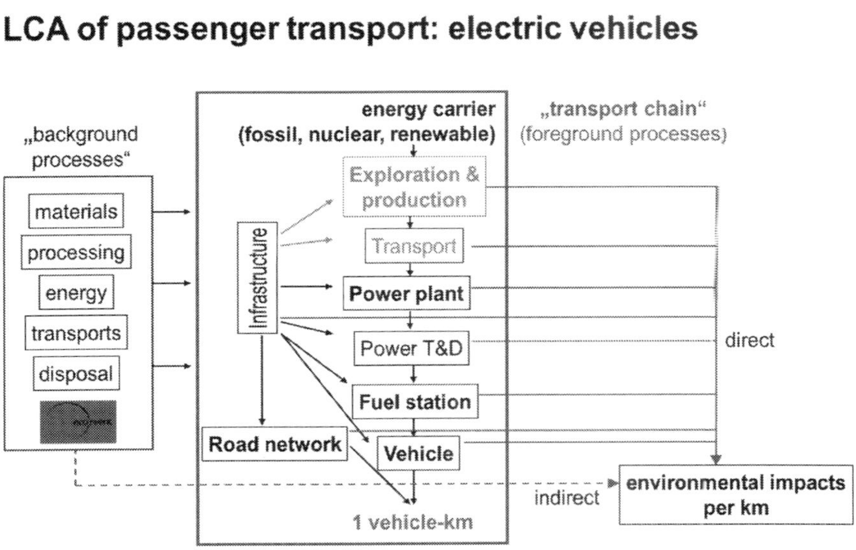

Figure 33: Product System for the LCA of Battery Electric Vehicles.

Source: Bauer, C., Hofer, J., Althaus, H.-J., Del Duce, A., Simons, A. (2015) The environmental performance of current and future passenger vehicles: Life cycle assessment based on a novel scenario analysis framework. Applied Energy, 157, 871–883, doi:10.1016/j.apenergy.2015.01.019.

To avoid shifting environmental impacts through the life cycle stages, we need to adopt and customize the life cycle approach. So, every industrial process should be carefully analyzed according to existing environmental and anthropogenic conditions and only after this Life Cycle Assessment can be applied.

Eco-Efficiency

The concept of eco-efficiency is important as a strategy for evaluating both ecology and economy in projects, activities and whole economies. Mostly eco-efficiency has been used on microscale to connect business profitability to environmental impacts of production processes as part of cleaner production initiatives and recently also connected to products (Garner and Keoleian, 1995).

Along with microlevel, there is also work going on at the macro-level and according to OECD the term 'eco-efficiency' as it was defined previously was insufficient on its own as a basis for policy making. We need wider understanding of the links between economic activity and environmental damage, driving forces of change and the psychological and ethical motives of producer and consumer behavior. Eco-efficiency is therefore expressed as the efficiency with which ecological resources are used to meet human needs (OECD, 1998).

The concept of eco-efficiency was first described by Schaltegger and Sturm (1989) and then widely publicized in 1992 in Changing Course (Schmidheiny, 1992), a publication of the World Business Council for Sustainable Development (WBCSD). Since then it has been accepted as the key strategic theme for global business in relation to commitments and activities directed at sustainable development. The WBCSD (1998) describes eco-efficiency as "being achieved by the delivery of competitively priced goods and services that satisfy human needs and bring quality of life, while progressively reducing ecological impacts and resource intensity throughout the life cycle, to a level at least in line with the Earth's estimated carrying capacity".

The concept's practical and theoretical importance lies in its ability to combine performance along two of the three axes of sustainable development, environment and economics. Issues concerning equity and other social properties are not included in the concept of eco-efficiency.

The notion that increasing economic development would have to be correlated with lowering of environmental impact is not a new concept (Ehrenfeld, 2005).

This relationship was first popularized through the socalled IPAT identity. Invented by Ehrlich and Holdren in the early 1970s, the relationship implies that to hold or reduce environmental impact (I) as population (P) and affluence (A) increase, technology (T) would have to become dematerialized (Ehrlich and Holdren, 1972). Later, environmental scholars spoke of Factor 4, Factor 10, or more as estimates of just how much dematerialization would be needed to offset economic and population growth (Reijnders, 1998). This broad description conveys the general thrust of the concept but offers no specifics as to how to reduce it to practice.

The First International Conference on Quantified Ecoefficiency (Leiden), in 2004, defined Eco-Efficiency fundamentally as "a ratio of some measure of economic value added to some measure of environmental impact. The higher the value added, the more efficient is the use of environmental services. Alternately, some invert the ratio, which then generally becomes known as eco-intensity. Marginal values may be used to determine relative performance among alternatives".

John Ehrenfeld pointed out the problem with quantification. He uses quantification broadly as the process of choosing and enumerating the expressions that will become the numerator and denominator of the ratio. Both are problematic. Computing the denominator (environmental impact) faces the same issues that are common to life-cycle assessment or any other environmental assessment methodology: What impacts should be included, what boundaries are appropriate, what output metrics should be used (mass, effects measures, values), and so on (Ehrenfeld, 2005).

The economic expression is somewhat less controversial, although not without problems. Some are analytic and others normative. Because sustainable development is about both human welfare and justice, one needs to ask whether a measure that is limited to the neoclassical notion of aggregate social welfare is consistent with this overarching context. Even if the answer is yes, the question of the validity of standard measures of welfare remains. How should the contribution of a firm be computed? The usefulness and acceptability of eco-efficiency is now and likely will continue to be buffeted by the same arguments that have affected the acceptability and legitimacy of both the numerator and denominator. This issue cannot be resolved without examining the way the results of eco-efficiency calculations are interpreted and applied in practice (Ehrenfeld, 2005).

Scientists pointed to several potential possibilities to resolve the problem of quantification:

- Choosing among alternative processes and products (microscale);
- Evaluating the performance of a company or other organizational entity;
- Evaluating the performance of a country, region, or other macro entity (Ehrenfeld, 2005).

For example, the chemical product company, BASF, uses a simple form of eco-efficiency to choose among alternative processes and options. Andreas Bode and his colleagues from BASF company argued that improving a process or even innovating a new or better one is a significant challenge for engineers. Engineers and production managers need to understand if and why process improvements or innovations are required and how dedicated process(es) are embedded in supply chain structures and plant interactions. They furthermore need to decide when their own assessment of improvement options has reached a level of reliability that allows the involvement of other management functions leading to more formalized financial and risk assessment procedures and detailed planning (Bode, A., et.al., 2011).

BASF's analysis of process improvement and eco-efficiency potentials is a three-stage process:

- The first stage features a production cost analysis of a process that needs improvement or faces replacement. The most important cost items and levers for cost reduction are identified.
- The second stage asks for the ideal process and distinguishes two types of potential improvement: ideal process technology (i.e. zero waste, no raw material excess, etc.) and ideal chemical reaction (i.e. no by-products and minimal possible raw material usage). The latter implies more radical changes and is required if technology improvements alone are insufficient.

- The third stage assesses these improvement options in terms of material and energy flows and related costs. BASF uses Umberto software to conduct these assessments. If the assessment results are promising, the improvement options are proposed as investment options and fed into formal financial appraisal and detailed planning processes (Bode, A., et.al., 2011).

BASF report exemplifies the eco-efficiency approach using a typical chemical production unit in which two raw materials and an inert material are converted into a main product, additional by-products and waste streams (Figurer 34). The unit consists of several processes, like synthesis, crystallization, separation, drying, milling, and dissolving that also involve material loops (Bode, A., et.al., 2011).

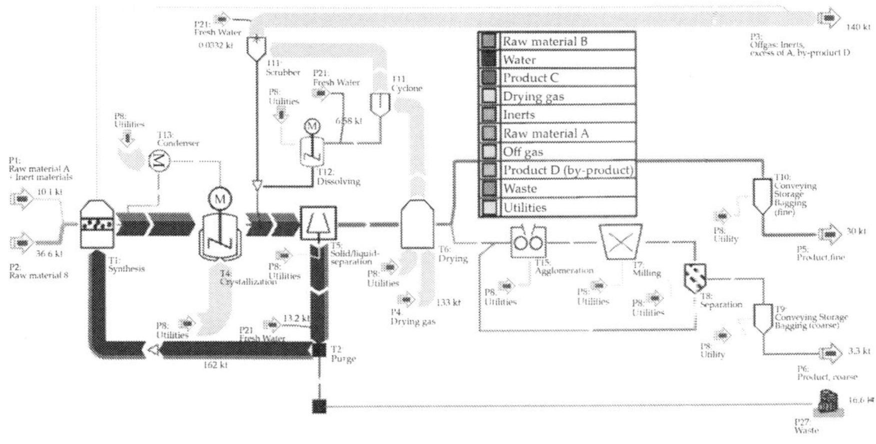

Figure 34: Sankey Chart of Mass and Utility Flows Prior to Process Improvement, BASF SE

Source: © BASF SE; Reproduction by courtesy of BASF SE

The production unit was considered no longer profitable due to its small contribution margin of about 2.5%. Hence a three-stage process improvement assessment was conducted to reduce production costs and increase the eco-efficiency, and the following findings were obtained:

- The production cost analysis revealed a rather typical manufacturing cost structure with no significant deviation from other processes and no outstanding cost component which could be tackled with priority. Not unexpectedly, the highest proportion of production cost was material cost.

- The evaluation of ideal processes revealed promising improvement potentials for both technology optimization as well as alternative chemistry, leading to potential production cost reductions of 10% and 30% respectively (Bode, A., et.al., 2011).

Figure 34 visualizes the status quo of the unit's material flows and Figure 35 reveals the related annual production costs.

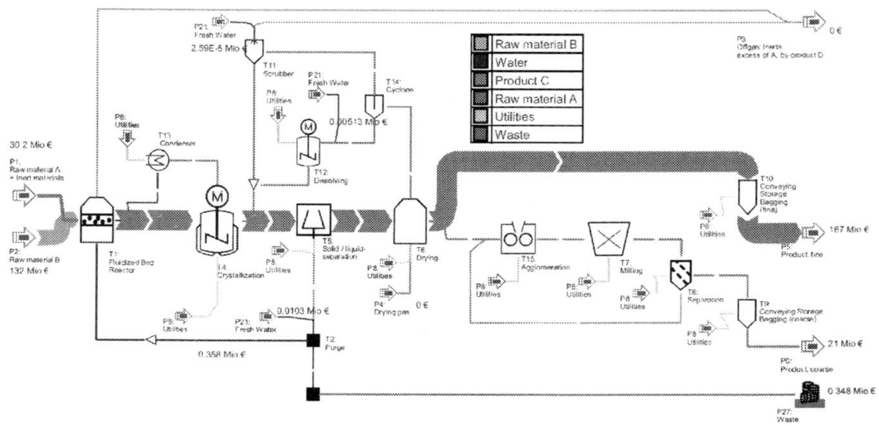

Figure 35: Sankey Chart of Material Flow Based Costs Prior to Process Improvement.

Source: © BASF SE; Reproduction by courtesy of BASF SE

Analysis of these diagrams along with the underlying process specifications and cause and effect relationships led to several suggestions for improvements. The most promising one was the idea of recovering product losses from the purge stream that was being treated as waste. This idea is depicted in the cost diagram in Figure 36 where we can notice the additional stream in the lower left corner that recycles some former product loss back to synthesis (Bode, A., et.al., 2011).

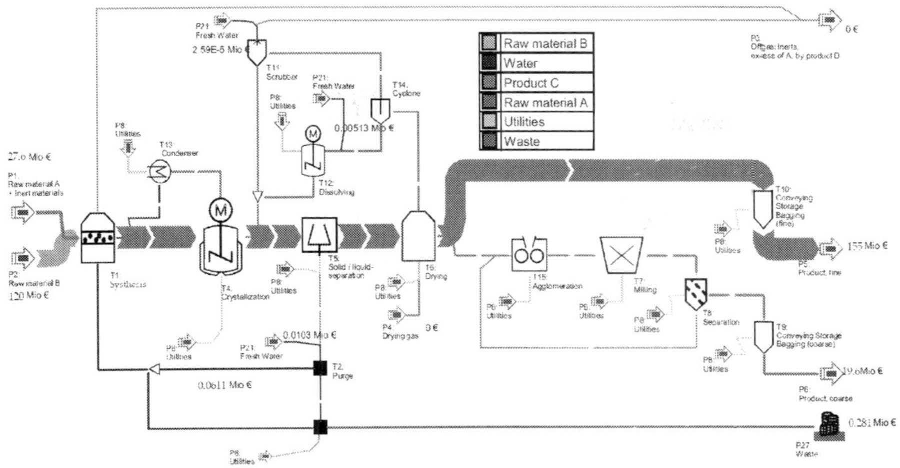

Figure 36: Sankey Chart of Material Flow Based Costs After Process Improvement.

Source: © BASF SE; Reproduction by courtesy of BASF SE

As we can analyze, application of eco-efficiency concept allowed BASF company to reduce raw material costs substantially (compare P1 and P2 on the left side of Figure 34 and 36) and down hence production and even waste treatment costs as well (compare P5, P6, and P27 on the right side of both figures). The three-stage eco-efficiency assessment helped BASF identify options to increase the contribution margin within that particular unit to competitive levels. Additionally, BASF experience reveals lower demand for raw material consumption, which in turn indicates higher resource efficiency during the production process, as well as smaller carbon footprints and less environmental impacts within the supply chain. Obtained results highlight the need for an integrated approach to material flow modeling and analysis and eco-efficiency (Bode, A., et.al., 2011).

At the macrolevel applications, eco-efficiency can be used to examine alternative governmental policies in the same way that companies rate product alternatives. It can also be used to characterize the performance of an economy by adding up the aggregate social welfare or value added and dividing by the total environmental impact. This procedure may be useful in comparing one nation or company to another, but again tells little about the direction of progress toward the goal of sustainable development (Ehrenfeld, 2005).

John Ehrenfeld (2005), the Executive Director of the International Society for Industrial Ecology, raised a very important question: ceo-efficiency is a notion that is meaningful only in the context of the economic model of sustainable development. Although the WBCSD's statements indicate that production output should be kept "in line with the Earth's carrying capacity," there is nothing in the analytic representation of eco-efficiency that provides a clue to this. Standard economic theory assumes that limitless resources will always be available as scarcity incentivizes sufficient innovation to produce substitutes. Carrying capacity simply does not enter the economic calculus. The present, clearly unsustainable state of the world trumps this theory, rendering eco-efficiency only a partially useful concept. Notwithstanding this limitation, more eco-efficient choices are always to be preferred over poorer choices, and in this sense eco-efficiency can be a useful tool for strategists and policy makers. On the other hand, its casual use can become seen as the "solution" to unsustainability (as business seems to be saying), shifting the burden (in the language of systems dynamics) away from the search for effective solutions. Better choices among a set none of which are good enough are fools' choices. Maintaining sustainability, for example, in terms of not exceeding carrying capacity, in an absolute, global sense is the goal, even of a process model such as sustainable development, in which this explicit goal is implicit (Ehrenfeld, 2005).

"If eco-efficiency is to become a useful indicator for determining choice, it must be coupled with other indicators and tools. Even the direction of change toward sustainability would be negative if increases in eco-efficiency were to be offset by larger increases in total output. It seems much more compelling that policy and product innovation should seek substitutes that radically reduce the amount of stuff that humans all over the globe use to produce well-being", concluded Ehrenfeld (2005).

Input-Output Economics

The growing share of trade flows in economic activities proved that production and consumption processes increasingly occur in different regions, sometimes located far apart one from each other. Input-Output Economics is the study of the interdependence of the different parts of an economic system. Quantifying the energy, materials, and emissions embodied in international trade has largely drawn on the Industrial Ecology concept of input-output analysis (IOA). Developed by economist Wassily Leontief to understand the interactions between economic sectors, IOA evaluates the economic and environmental implications of changes in final demand on the structure of economic activities (Leontief, 1970).

Among his colleagues were Walter Isard, who contributed substantially to the initial input–output analysis at the regional level (Isard & Cumberland, 1961). The International Input-Output Association was created only in 1988 and its journal, Economic Systems Research, dates to 1989, the community has been holding international conferences since 1950 (Duchina and Hertwichb, 2003). Other developments have included multiregional input–output tables (MR-IOT), substituting import and export vectors with full tables for multiple regions (Wiedmann et.al., 2007). Simultaneously, other factor inputs, such as land, materials and water consumption, as well as air emissions, were quantified to expand the analysis of material flows and compile indicators. such as the ecological footprint of nations (Yamano and Ahmad, 2006).

The input–output framework as exemplified by the European System of Accounts (ESA95) consists of three types of table: supply and use tables (SUT) and symmetric input–output tables (IOT). The supply table shows the supply of goods and services, both domestic and imported, by product and type of supplier in basic prices, while the use table shows the use of goods and services by product and type of use in purchase prices, i.e. as intermediate consumption by industries, final use (consumption, gross capital formation) and exports. The use table also contains the components of the value added by industry, i.e. compensation of employees, other taxes less subsidies on production and gross operating surplus. The use table can be converted to basic prices with the help of valuation matrices reflecting retail, wholesale and taxes/subsidies per product used per industry. If necessary, the SUT can be broken down into a domestic and import (use) and an export (supply) part (European Communities, 1996; UN et al., 2003).

Input-output tables (IOTs), or matrices, detail the supply and use of products (goods and services) sector by sector, including imports and exports and final consumption categories such as households, government spending, and capital formation. What is true for economic sectors is also true at the process level, and supply and use tables essentially represent the transfer coefficients from one industry to another in monetary or physical terms. The availability of environmentally extended input–output tables (EE-IOT) has considerably increased the usefulness of IOTs in MFA at the level of industrial activities Environmental accounts typically include energy intensity and carbon emissions by sector, which is especially relevant for economic and environmental policies. Because IOA is linear in nature, the impacts of changes in the final demand can easily be estimated for specific industrial sectors or on the structure of economic activities—that is, shifts in production domestically and abroad (Duchina and Hertwichb, 2003).

Today statistical offices around the world compile and publish input-output tables and related data on a regular basis as a central part of their national accounts (NEDA, 2011). NEDA and JICA developed the Training Kit for the Basic Training in IO Analysis, which is based on the three types of tables (Figure 37)

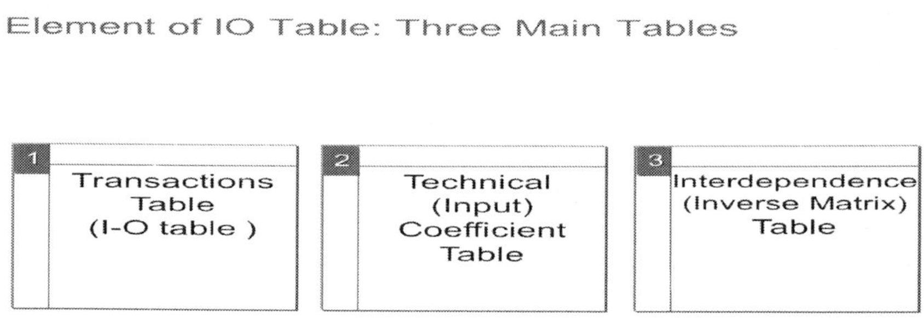

Figure 37: Three Elements of IO Analysis.

Source: Source: JICA-PEDPIO, JICA and NEDA, February 2011. Available at: http://w3.neda.gov.ph/iosimulator/3.html

The input-output transactions table, derived from economic censuses and surveys, describes all economic transactions in an economy with an especially detailed focus on inter-industry purchases and sales as it is shown at the Figure 38 (NEDA, 2011).

Transactions Table (Philippine I-O table)

3-Sector Input-Output Transactions Table (Competitive Type) [2006]

(In Million Pesos at Current Producer's Price)

			Intermediate Demand				Final Demand				Imports	TOTAL GROSS OUTPUT
			AGRICULTURE, FISHERY & FORESTRY	INDUSTRY	SERVICES	TOTAL INTERMEDIATE DEMAND	CON-SUMPTION	INVESTMENT	Exports	TOTAL FINAL DEMAND		
			I	II	III	TID				TFD	M	TGO
Intermediate Input	I	AGRICULTURE FISHERY & FORESTRY	34,373	665,978	28,151	728,502	297,687	98,665	26,799	372,433	(50,713)	1,100,935
	II	INDUSTRY	132,264	3,513,302	1,064,629	4,710,194	2,191,621	880,429	2,041,711	2,299,759	(2,814,0■3)	7,009,953
	III	SERVICES	144,462	660,259	1,003,064	1,807,785	2,616,225	99,862	785,028	3,358,972	(142,143)	5,166,757
	TII	TOTAL INTERMEDIATE INPUTS	311,099	4,839,538	2,095,844	7,246,481	5,105,532	1,078,956	2,853,538	6,031,164	(3,006,363)	13,227,645
Value Added	V1	WAGES & SALARIES	210.545	558,780	1,135,553	1,904,878						
	V2	INDIRECT TAXES LESS SUBSIDIES	26,944	89.915	116,589	233,448						
	V3	DEPRECIATION	74,020	219,448	288,318	581,787						
	V4	OPERATING SURPLUS	478,327	1,302,272	1,530,453	3,311,052						
	TP1	TOTAL PRIMARY INPUTS (GVA)	789,836	2,170,415	3,070,913	6,031,164						
	TGI	TOTAL GROSS INPUT (GROSS OUTPUT)	1,100,935	7,009,953	5,166,757	13,277,645						

3 Sectors

Figure 38: Example of Transaction Table (Phillippine I-O Table).

Source: JICA-PEDPIO, JICA and NEDA, February 2011. Available at: http://w3.neda.gov.ph/iosimulator/3.html

The second element of IOT table, Technician Coefficients, is depictured at the Figure 39. An individual balance equation in the physical input-output model assures that those outputs of a sector distributed to other sectors and those going to consumers add up to total production. Corresponding to the physical model is a price model: a balance equation in the price model assures that the costs of individual inputs to production plus profits add up to the unit price (NEDA, 2011).

Technical Coefficients Table (Philippine I-O table) 2

Intermediate Input Coefficient (**A** Matrix)

		1	2	3
1	AGRICULTURE, FISHERY & FORESTRY	0.031	0.095	0.005
2	INDUSTRY	0.120	0.501	0.206
3	SERVICES	0.131	0.094	0.194
Total Intermediate Inputs		0.283	0.690	0.406
V	Wages and salaries	0.191	0.080	0.220
	Indirect taxes less subsidies	0.024	0.013	0.023
	Depreciation	0.067	0.031	0.056
	Operating surplus	0.434	0.186	0.296
Total Gross Value Added [NIT+OVA]]		0.717	0.310	0.594
TOTAL INPUTS (=TOTAL OUTPUTS)		1.000	1.000	1.000

Primary Input Coefficient (**GVA** Rate)

Figure 39: Example of Technician Coefficient Table.

Source: JICA-PEDPIO, JICA and NEDA, February 2011. Available at: http://w3.neda.gov.ph/iosimulator/3.html

The equations are solved for total – direct plus indirect – production requirements to satisfy given deliveries to consumers and for the vector of prices (or price deflators in the common case where the variables in the physical model are measured in money units) that covers both intermediate inputs and costs of factors of production. It is converted by a simple 6 manipulation to a matrix of inputs per unit of output, which serve as structural parameters. This is a crucial step because it permits the passage from description and descriptive statistics to modeling of alternative scenarios (Moreau and Massard, 2017). The matrix and other related data are manipulated in a model that, in its simplest static form, is a system of balance equations involving the matrix inverse, sometimes called the Leontief inverse (Figure 40) (NEDA, 2011).

Interdependence Coefficient (Inverse Matrix) Table 3

✓ The inverse matrix $(I-A)^{-1}$ is fundamental to input-output analysis as it shows the full impact of an exogenous increase in final demand on all industries.

		1. AGRICULTURE, FISHERY & FORESTRY	2. INDUSTRY	3. SERVICES	Total
1	AGRICULTURE, FISHERY & FORESTRY	1.0460	0.1436	0.0300	1.2195
2	INDUSTRY	0.1530	1.5242	0.2479	1.9250
3	SERVICES	0.1809	0.1937	1.2641	1.6388
	Total (Output Multiplier)	1.3799	1.8615	1.5420	1.5944 Ave. of sector1~3

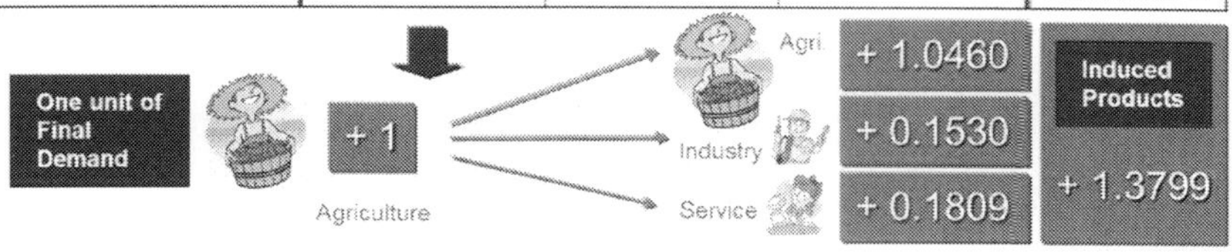

Figure 40: Example of Inverse Matrix Table.

Source: JICA-PEDPIO, JICA and NEDA, February 2011. Available at: http://w3.neda.gov.ph/iosimulator/3.html

A column in an input-output matrix is a compact representation of the direct inputs to production for a given sector. The corresponding column of the inverse matrix captures the indirect inputs from all other sectors as well as the direct ones. It is this feature which is of special interest for life cycle analyses. While the input-output table and the static physical input-output model have been used successfully and extensively in LCAs, input-output economists also make use of a variety of conceptual extensions including the static price model (mentioned earlier), dynamic physical and price models, and optimization models that select low-cost technologies. These models are used with data collected in a variety of ways to explore not only the impact of actual or hypothetical products and processes but also to evaluate the implications of moderately detailed scenarios about the future (Moreau and Massard, 2017).

Alternative scenarios about the future play a role in input-output analysis that is similar to that of alternative actual or hypothetical products and processes in an LCA study, and collaboration between these communities is now established and poised to expand. Recently a joint Working Group between the International Society for Industrial Ecology and SETAC was created on Input-Output Analysis in Life Cycle Assessment. One area where collaboration may prove especially fruitful is in bringing household lifestyles and consumption into the purview of Industrial Ecology. There is a substantial and growing body of research on this subject (Duchina and Hertwichb, 2003).

José Manuel Rueda-Cantuche and his colleagues (2009) with the financial support of the EU's 6th Framework Programme developed EXIOPOL input-output analysis (EE I–O) software which organizes data from the very different sources in a similarly structured way. Figure 41 presents tools used in this input-output analysis.

Table 1- Tools needed for analysis of cost-effectiveness and cost -benefit assessment of sustainability options at micro, meso and macro level.

Level	Example	Tools for direct cost assessment	Tools for impact assessment	Tools for benefit assessment[a]	Gaps in toolbox filled by EXIOPOL
Micro	Company level (e.g. technology choice), household level (e.g. consumption decisions)	Life cycle costing	Life cycle assessment, site specific tools like Ela or RA	External costs, local	*Externalities*: external casts are still lacking for many prassures and end points, a range of conceptual questions have still to be answered, and cost data for ready use in maso-and macro-accounting frameworks are not available.
Meso	Sector or product group level (e.g. sector standard, tax, or ban)	IO, in combination with scenario tools (e.g. CGE)	EE I-O, in combination with scenario tools (e.g. CGE)	External costs, sectoral average	IO and EE I-O: comprehensive, detailed, uniform and open- source IOTs with environmental extensions for the EU27, and with suitable trade links with non-EU contries are absent. Such a
Macro	Societal level (as per sector level)	IO, in combination with scenario tools (e.g. CGE)	EE I-O, in combination with scenario tools (e.g. CGE)	External costs, country average	comprehensive EE I-O database can also support LCA and LCC calulations by providing the system context which allows the estimation of impacts of truncated processes etailed by LCA approaches [b].

Note: EIA: Environmental Impact Assessment. LCA: Life Cycle Assessment. LCC: Life Cycle Costing. IOTs: Input-Output Tables. RA: Risk assessment: CGE: Computable General Equilibrium (model).

[a] Note: we assume here that direct benefits can easily be included in direct (net) cost assessments, so that here we in fact have to focus on benefits in terms of the reduction of external costs.

[b] LCAs try to follow the entire supply chain to the final functionality delivered, but due to the networked character of any economy seemingly unimportant supply chains are at some point 'cut off' and neglected in further analysis (e.g. Guinee, 2002). The cumulative effects of such cut-offs can be up to 50% of the total impact of a product (Lenzen, 2001) which can be corrected by hybrid IOA-LCA (e.g. Heijungs and suh, 2002; Hendrickson et al., 2006).

Figure 41: Tools used in EXIOPOL database.

Source: Rueda-Cantuche et.al., (2009). Towards a global multi-regional environmentally extended input-output database. https://www.researchgate.net/publication/46490457_Towards_a_global_multi-regional_environmentally_extended_input-output_database [accessed May 3, 2017].

The principal algorithm of the EXIOPOL input-output analysis is shown at the Figure 42. The EE I–O database allows to analyze, monitor, and improve assessment of issues such as the total environmental impacts and external costs of each industrial sector, each final consumption activity and each final product group in relation to imports and exports and of each resource used (Rueda-Cantuche et.al., 2009).

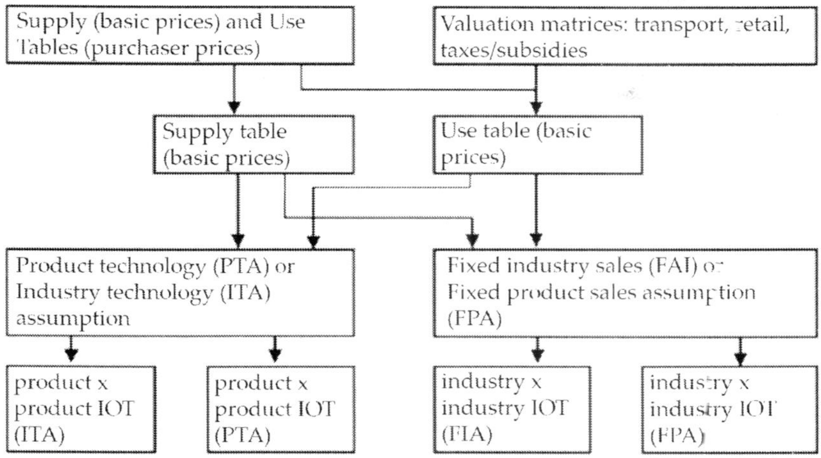

Figure 42: Simplified Input – Output Framework.

Source: Rueda-Cantuche et.al., (2009). Towards a global multi-regional environmentally extended input-output database. https://www.researchgate.net/publication/46490457_Towards_a_global_multi-regional_environmentally_extended_input-output_database [accessed May 3, 2017].

Project team used the latter route since it defers the need to make theoretical assumptions to arrive at an IOT until the final stage. Taking into account that most extensions (emissions and resource uses) were linked to industries which emit and use primary resources, trade links and analyses related to final demand require a product perspective, and that IOTs focus either on products or on industries and cannot deal with both perspectives at the same time, the choice of supply and use tables (SUTs) was made (Figure 43) Rueda-Cantuche et.al., (2009).

For each country, the following dataset was created:

- a supply table and a use table (the former is usually published in basic prices, the latter in purchase prices);
- valuation matrices (retail and wholesale, transport, taxes, and subsidies on products, which enable use tables in purchase prices to be converted to basic prices);

- a breakdown of the use table into a domestic use table and import use table, or alternatively a breakdown of the supply table into a domestic supply table and an export supply table (Rueda-Cantuche et.al., 2009).

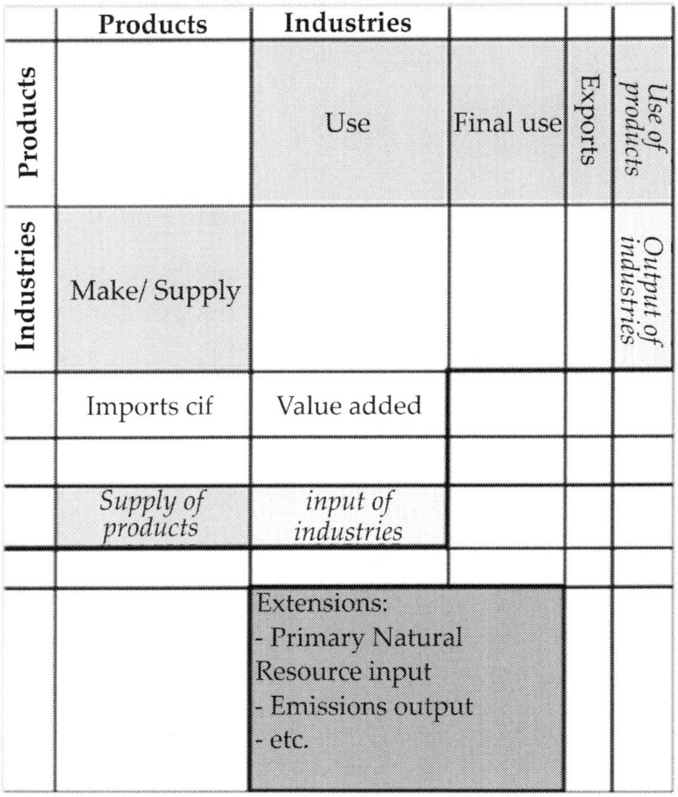

Figure 43: Schematic SUT with Environmental Extensions.

Source: Rueda-Cantuche et.al., (2009). Towards a global multi-regional environmentally extended input-output database. https://www.researchgate.net/publication/46490457_Towards_a_global_multi-regional_environmentally_extended_input-output_database

The choice of SUTs in the framework gives the design of the database a fundamental advantage with regard to developments in the standardization of data collection. Database allows to estimate some of the SUT data due to data and time constraints, the content of the database can be updated and improved over time without any need to change the structure. The tool allows to make the structural path analysis and contribution analysis (i.e. which sectors or processes contribute to what extent to impacts or external costs related to products or resources used, and where these processes are located. It uses a number of widely used and inspiring indicators: external costs, ecological footprint, and various LCA impact categories. Additionally, the EXIOPOL database supports the calculation of the Ecological Footprint, thus, the project can help in the further dissemination of that indicator (Rueda-Cantuche et al., 2009).

One of the main advantages of EE-IOTs for country level is to provide a perspective on material and energy flows, accounting for environmental effects at all levels in the process of production and consumption. Input-Output Analysis has also been combined with Life-Cycle Assessment for inventories for which process data is lacking or to improve overall data reliability; in turn, the results of Input-Output Analysis can inform macroeconomic policies as well as microeconomic consumer choices Rueda-Cantuche et.al., (2009).

Choosing the Industrial Ecology concepts and tools to use in practical tasks, we should remember that every manufacturing and environmental case is unique, and hence it is very difficult to come up with some universal design principles for or define 'conditions' of a successful industrial ecosystem for 'ideal' industrial ecosystem management. The important point is to be aware of some of the main conditions of inclusive and sustainable industrial development.

4 CHAPTER

CLASSIFICATION OF INDUSTRIAL ENVIRONMENTAL IMPACT

CLASSIFICATION APPROACHES

Talking about industrial environmental impact we must remember that there no 'ecologically clean' technology in the world as some manufacturers advertise in their business papers. Every human technology impacts the planet environment and planet Earth, in return, gives its answer to the humanity in the form of retroaction. We must remember: every environmental impact eventually will affect people.

Industrial environmental impact can be classified from various points. First of all, environmental impacts fall into three categories:

- direct impacts;
- indirect impacts;
- cumulative impacts.

These three types or categories of impact must be considered during the National Environmental Policy Act process in the USA (WSDOT, 2015). Table 3 provides a summary comparison of direct, indirect and cumulative effects.

Table 3: Comparison of Direct, Indirect, and Cumulative Effects

Type of Effect	Direct	Indirect	Cumulative
Nature of effect	Typical/inevitable/ predictable	Reasonably fore-seeable/probable	Reasonably foreseeable/ probable
Cause of effect	Project	Project's direct and indirect effects	Project's direct and indi-rect effects and effects of other activities
Timing of effect	Project construction and implementation	At some future time after direct effects*	Past, present, or in the future
Location of effect	Within project im-pact area	Within boundaries of systems affected by project	Within boundaries of systems affected by the project

*Indirect could potentially occur before the project is built (i.e., speculators initiating land use actions in anticipation of project construction).

Source: A Guidebook for Evaluating the Indirect Land Use and Growth Impacts of Highway Improvements, Final Report SPR 327, Oregon DOT and FHWA, April 2001.

Direct impacts. Direct impacts occur through direct interaction of an activity with an environmental, social, or economic component. The technological actions cause direct impacts, for example, during open mining process there are such direct impacts as land consumption, removal of vegetation, ground corruption, air pollution, etc. When Wastewater Treatment Plant dumps insufficiently treated waters into a river, it may lead to a decline in water quality in terms of high biological oxygen demand or dissolved oxygen or rise of water toxins. Direct environmental influence is easier to account, assess, and control than indirect impacts, because the cause-effect relationship is understandable.

Indirect impacts. Indirect impacts on the environment are these which doesn't cause the direct damage, often produced away from a place of activity or because of a complex impact pathway. World Bank book on environmental effects classification describes indirect impacts as "secondary, tertiary, and chain impacts and they may have more profound consequences on the environment and on the human health than direct impacts. Indirect impacts are more difficult to measure, but can ultimately be more important. Over time they can affect larger geographical areas of the environment than anticipated" (World Bank, 1998).

For example, sulfur oxide in form of SO2, emitted to the air by vehicles, may deposit on land as SO4 and cause acidic soils. Another example of indirect impact is the decline in water quality due to rise of water bodies temperature caused by receiving cooling water discharge from the nearby power plant. This may, in turn, lead to a secondary indirect impact on aquatic flora in that water body and may further cause reduction in fish population. Reduction in fishing harvests, affecting the income of fishermen is a third level impact. Such impacts are characterized as socioeconomic (third level) impacts.

Environmental impacts should be considered not only as they pertain to manufacturing process, but also to sites associated with the whole production chain. For example, electricity production process at the Thermal Power Plant includes environmental effects from mining fossil fuels, transporting and refining them, impacts from waste deposits, materials treatment fields, roads, and facilities provided for plant workers. These 'off-row' areas should be predicted and evaluated by taking an integrated approach to impact assessment.

It is especially important that any synergetic relationships between impacts be closely examined, since indirect effects frequently lead to synergetic impacts. It is with indirect effects that impact linkages between the natural and social environment often take place. For example, the appropriation of land to build a road may displace farmers, and may interfere with their cropping pattern and force them to use another water supply. This change could result in a depletion of a groundwater aquifer, intensification of new land clearing, erosion, water runoff contamination with added fertilizers and pesticides, etc. (World Bank, 1998).

Proper documentation of the assumptions used in the impact report is critical. Assessing any impacts that are difficult to quantify, error limits and the quality of basic information must be indicated.

Cumulative impact. Cumulative impact consists of an impact that is created as a result of the combination of the particular manufacturing project together with other projects causing related impacts. These impacts occur when the incremental impact of the project is combined with the cumulative effects of other past, present, and reasonably foreseeable future projects. Cumulative environmental changes could be caused by such types of anthropogenic actions:

- single large events;
- multiple interrelated events;
- catastrophic sudden events;
- incremental, widespread, slow change.

These can generate additive, multiplicative or synergetic effects, which can then result in damage to the function of one or several ecosystems (such as the impairment of the water regulation and filtering capacity of a wetland system by construction of a road across it), or the structure of an ecosystem (such as placement of a new road through a forest, leading to m-migration or land clearing which results in severe structural loss to the forest) (World Bank, 1998).

Cumulative effects can be evaluated through the Cumulative Effects Assessment – a complex process which requires extensive knowledge of ecological principles and ecosystem response mechanisms. The success of a cumulative impact evaluation relies on the assessment framework that is set up before the evaluating process.

The cumulative impacts can be, due to induced actions of projects and activities that may occur if the action under assessment is implemented such as growth inducing impacts and other effects related to induced changes to the pattern of future land use or additional road network, population density or growth rate. Induced actions may not be officially announced or be part of any official plan. Induced effects usually have no direct relationship with the action under assessment and represent the growth- inducing potential of an action (WSDOT, 2015).

However, the cumulative impacts due to induced development or third level or even secondary indirect impacts are difficult to be quantified. Because of higher levels of uncertainties, these impacts cannot be normally assessed over a long-time horizon. An environmental impact assessment practitioner usually can only guess as to what such induced impacts may be and the possible extent of their implications on the environmental factors (WSDOT, 2015). World Bank Handbook on types of environmental effects proposed such an analysis of cumulative impacts on the example of stream (Figure 44).

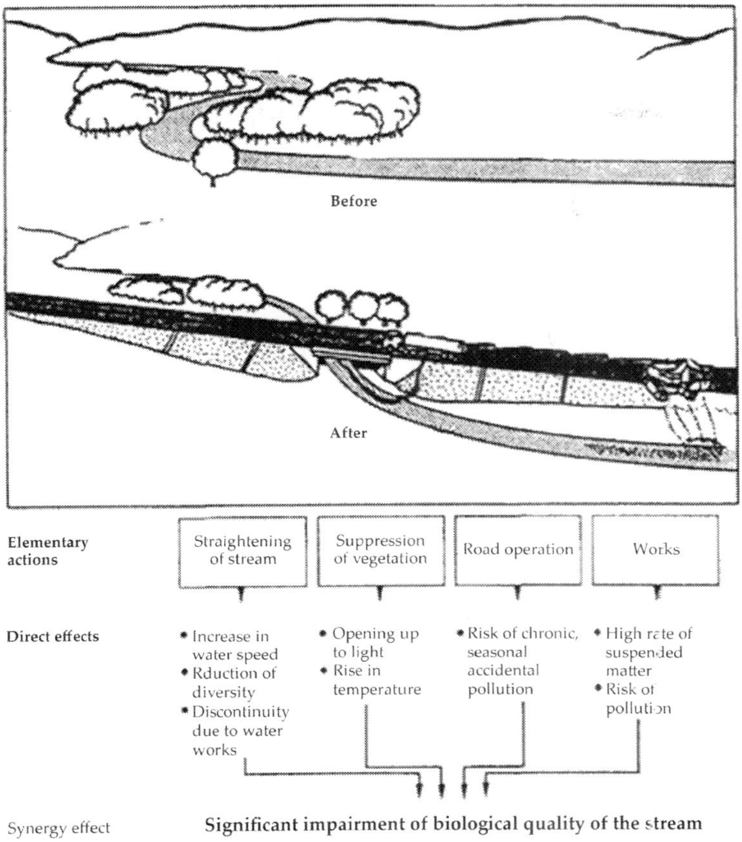

Figure 44: Cumulative Impacts: The Example of a Stream.

Source: World Bank (1998). Types of Environmental Impact. In: Roads and The Environment: A Handbook. Washington, DC, World Bank, Environment Department, 60-65.

One of the technical subset or variant of cumulative impact is the ecosystem function impact. Ecosystem function impacts disable or destabilize whole ecosystems and they are the most dangerous and often the least likely to manifest themselves over a short period of time (World Bank, 1998).

One striking example is the highway built across a mangrove forest, 100 ha in size, at the Caribbean coast. At the planning stage to designers didn't fully understood what extent the fresh and sea water needed to mix in order for the healthy forest to function on both sides of the road. After road was built, most of the forest has died off. The reason was that on one side of the road the waters were not saline enough, and on the other there was not enough mixing with fresh water. Influenced by change of water salinity, the mangrove trees, some over 20 m high, were destroyed by other factors (disease, etc.). The effect on the ecosystem was exhausting. Local population which used that mangrove forest for living also was impacted. What was remarcable, no sign of this impact appeared until two to three years after the road was built (World Bank, 1998).

World Bank handbook on environmental effects types brakes direct, indirect, and cumulative environmental impacts further down according to their nature, into:

- random and predictable impacts;
- local and widespread impacts;
- temporary and permanent impacts;
- short- and long-term impacts;
- positive and negative impacts (World Bank, 1998).

Random / predictable impacts. We should differentiate assured, or highly probable impacts, and more random, or unpredictable ones, which have a low occurring probability but may have serious consequences for the environment, during the preliminary analysis of an environmental impact assessment.

For example, in a region with a large, densely settled population, it is reasonable to predict that the construction of an oil treatment plant will result in air pollution, whereas incidents such as accidental pollution, fire, or spillage of toxic products are, by nature, unpredictable. Well understood and predictable impacts can usually be mitigated with remedial measures, and therefore call for minor environmental assessment requirements such as an IEE and environmental summary report, as opposed to a full environmental assessment (World Bank, 1998).

Local / widespread impacts. Local impacts include effects in the immediate local area of production process. Widespread impacts can occur many kilometers from the project. These impacts are often linked to indirect effects that arise over the medium or long-term existence of the project and can include as examples the influx of settlers, deforestation, and the development of new industries (World Bank, 1998). The scale of considering the impact local or widespread depends on the scale and type of manufacturing process that is under evaluation. For the impact assessment of local dairy producer, the area of local effects is smaller than for metallurgy enterprise.

According to World bank (1998), while the focus of most local effects has been on relatively narrow corridors and circles around the plants, impacts can extend much further. Major habitat conversion can take place up to 10-70 km in diameter of the industrial activity. For example, air pollution from cement production could spread on the territory up to 50 km in diameter.

Temporary / permanent impacts. Temporary impacts are those whose occurrence is not lasting, and which will eventually reverse themselves, the affected system having returned to its previous state (World Bank, 1998). Good example of such an effect is the trampling vegetation during rising the wind power turbine; it recovers after a few months, to the point where no change from the original state is noticeable.

Permanent impacts are those which are irreversible – the affected system will not return to its previous state on a human timescale. It is important to note that permanent from the viewpoint of environmental assessment, is defined as 'within one's lifetime' (World Bank, 1998). Therefore, the burning and destruction of a tropical forest in Indonesia for the needs of mining industry would be considered as permanent impact.

Short- / long-term impacts. Short-term impacts are those which appear during or shortly after manufacturing process or some construction; long-term impacts may arise during construction, but many of their consequences appear during the operational phase, and may last for decades (World Bank, 1998).

Positive / negative impacts. The emphasis of Industrial Ecology is primarily on avoiding and mitigating negative impacts, which should be a high priority in all industrial projects. World Bank (1998) admitted that sometimes environmental impacts have both positive and negative effects; some impacts can positively affect some people and negatively affect others in the same environment. Positive outcomes that occur because of project completion typically include improved access, reduced travel time and cost, and perhaps reductions in accidents or noise for people. Other positive outcomes can be designed into a project, for example, improving water retention for local use, flood control, or providing better facilities for local community. In some cases, positive impacts can appear without having been initially foreseen by the road agency, such as the use of borrow sites to water livestock in dry areas (World Bank, 1998).

To qualify environmental impacts by the type of effect they have on the environment is not sufficient. World Bank Manual (1998) recommends that impacts must also be categorized according to their serenity / seriousness. The most damaging and longest lasting impacts should be the first to be avoided or mitigated. Defense National Environmental Standard developed the Guidance for Assessment of Environmental Incident Severity (refer Table 4).

Table 4: Guidance for the Initial Assessment of Environmental Incident Severity

Major	Moderate	Minor
Significant AND irreversible impact on the environment. OR Environmental impact requires emergency commitment of significant time and/or money to remediate, OR would take more than 10 years to recover naturally. OR National / international damage to Defense's reputation. Negative press coverage at a national or international scale.	Significant but reversible, OR irreversible and localized, impact on the environment. OR Environmental impact requires a programmed commitment of moderate time and/or money to remediate, OR would take 2-10 years to recover naturally. OR Local / regional damage to Defense's reputation. Negative press coverage at a local or regional scale.	Reversible and localized impact on the environment. OR Environmental impact requires a programmed commitment of some time and / or money to remediate, OR would take less than 2 years to recover naturally. OR Damage to Defense's reputation with a single or small group of organizations or individuals.

Source: Defense National Environmental Standard (DNES), (2014). Available at: http://www.defence.gov.au/estatemanagement/lifecycle/IncidentManagement/Environment/Docs/DNESScheduleAGuidanceforAssessmentofEnvironmentalIncidentSeverity.pdf

Proposed guidance should be used in the initial assessment of the severity of an environmental impact or incident. 'The environment' is defined as "Natural surroundings in which Defense personnel and its contractors operate, including air, water, land, natural resources, flora, fauna, people and their interrelation. For the purposes of this Defense National Environmental Standard and to ensure Defense meets its legislative obligations, 'environment' also includes indigenous, historic, and natural heritage" (DNES, 2014).

Environmental workers should use the Guidance for the Assessment of Environmental Incident Severity to categorize the seriousness of an environmental incident (refer Appendix A). The column 'incident category' shows us receptors and activities as both are relevant in determining the severity of an incident. An environmental accident may fall into more than one incident category; in that case, the severity should be assessed for all relevant incident categories and the highest rating should be used to classify the incident. For example, an environmental incident may occur which is categorised as minor under the 'General' incident category. However, if the incident is related to an issue of interest to local or regional media and the 'Reputation' rating is moderate, the incident's overall rating would be moderate (DNES, 2014).

Talking about types and serenity of environmental impacts, we must consider Ecosystem Function Impacts and Synergy Impacts.

Ecosystem Function Impacts. Ecosystem function impacts are technically a subset or variant of cumulative impacts, the most dangerous and often the least likely to manifest themselves over a short period of time. Such impacts usually disable or destabilize whole ecosystems, (World Bank, 1998).

The impact of biodiversity on any single ecosystem process is nonlinear and saturating, such that change accelerates as biodiversity loss increases. The form of biodiversity–ecosystem functioning indicates that initial losses of biodiversity in diverse ecosystems have relatively small impacts on ecosystem functions, but increasing losses lead to accelerating rates of change (Figure 45).

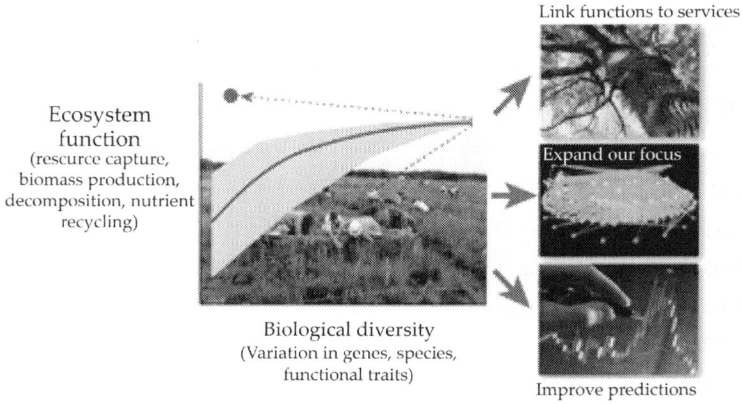

Figure 45: The Form of a Typical Diversity–Function Relationship.

Source: Cardinale, B.J. et.al., (2012). Biodiversity Loss and its Impact on Humanity. Review. Nature | VOL 486 | 7 June 2012. doi:10.1038/nature11148. Available at: http://snre.umich.edu/cardinale/wp-content/uploads/2012/04/Cardinale-et-al-Nature-2012.pdf

Conceptual diagram at Figure 45 summarizes what we know about the shape of the biodiversity-ecosystem functioning relationship. The red line shows the average change across all combinations of genes, species, or traits. The grey polygon represents the 95% confidence interval, whereas red dots give maximum and minimum values of the most or least productive species grown alone in monoculture. Diverse communities are more productive because they contain key species that have a large influence on productivity, and differences in functional traits among organisms increase total resource capture. Diversity effects are driven by single, highly productive species versus some form of 'complementarity' among species. Research and syntheses over the past 10 years have made it clear that both the identity and the diversity of organisms jointly control the functioning of ecosystems. Quantification of the variance explained by species identity versus diversity found that, on average across many ecosystems, each contributes roughly 50% to the net biodiversity effect. Complementarity may represent niche partitioning or positive species interactions, but the extent to which these mechanisms broadly contribute to ecosystem functioning has yet to be confirmed (Cardinale, B. J. et al., 2012).

One striking example of ecosystem function impact is the highway constructed across a mangrove forest (100 ha in size) along the Caribbean coast. It was not fully understood at the planning stage to what extent the fresh and sea water needed to mix in order for the healthy forest to survive on both sides of the road. As a result, most of the forest has died off; on one side the waters were not saline enough, and on the other there was not enough mixing with fresh water. The effect on the ecosystem was devastating and the impact on the local population which used the mangrove forest area was severe. Almost certainly, no sign of this impact appeared until two to three years after the road was built. This case could develop in situations where roads cross wildlife migration routes, that can impose stress on the migratory population for many generations, or even permanently, and cause instability, increased mortality, and possibly catastrophic decline (World Bank, 1998).

Ecosystem effects usually include impacts on social environment. Figure 46 shows the WWF infographics on humanity ecosystem function effects.

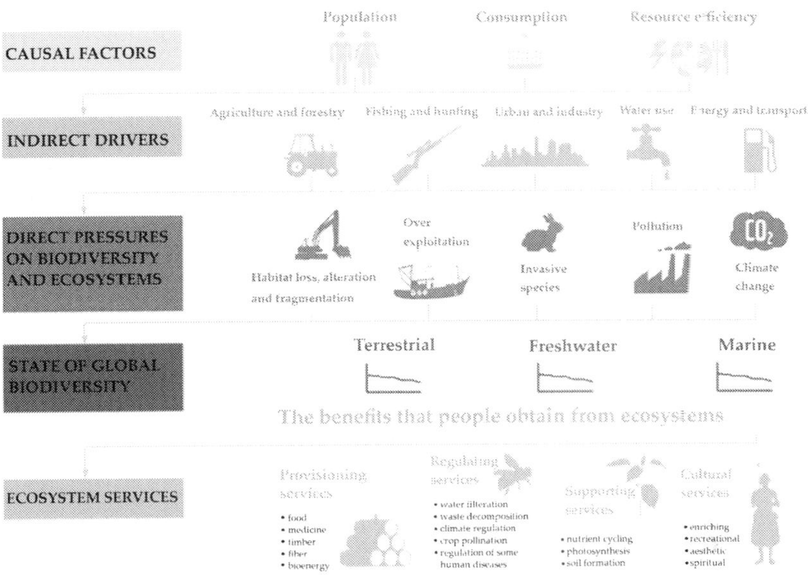

Figure 46: Humankind Impact on Ecosystem Functioning.

Source: WWF Living the Planet Report (2012).

The loss of natural capital and the ecosystem services affects the pollination of crops, the formation of fertile soil and the fast disappearing glaciers. In turn that will drastically reduce the flow of rivers, which will cause food and water shortage and affect every inhabitant of this planet, from the Himalaya to Europe (WWF, 2012).

Synergy effect: When environmental impact from some industrial process occurs, each elementary action produces a certain effect or a risk that can be limited, but the combination of such factors and therefore their consequences may be the source of significant synergy effect. In the synergistic effect, an interaction of two or more chemicals can results in an effect greater than the sum of their separate effects.

Mauderly and Samet pointed out such potential interactions among pollutants:

- Additivity: effect of the combination equals the sum of individual effects;
- Synergism: effect of the combination is greater than the sum of individual effects;
- Antagonism: effect of the combination is less than the sum of individual effects;
- Inhibition: a component having no effect reduces the effect of another component;
- Potentiation: a component having no effect increases the effect of another component;
- Masking: two components have opposite, cancelling effects such that no effect is observed from the combination (Mauderly and Samet, 2009).

'Effect' means the observed expression of the particular health outcome in question. A combination of pollutants could have different interactions for different outcomes. The interaction could occur at any level of biological pathway from exposure to expression of the outcome (Mauderly and Samet, 2009).

According to Mauderly and Samet (2009), synergism between two environmental pollutants can occur through different mechanisms. Pollutants can act in the same mechanistic pathway at the same or different steps; or the presence of one can influence the ability to mitigate the action of the other; or the presence of one can influence the dose of the other. Under the particular conditions, the existence of synergism could be dose dependent, that the same combined exposure might be synergistic for one effect and not for others, or that the same effect may be synergistic in some tissues and not in others (Mauderly and Samet, 2009). Analytical research made by Joe L. Mauderly and Jonathan M. Samet demonstrates synergy effects between O3 and other pollutants (refer Appendix B).

Synergisms among occupational inhalation exposures are known; perhaps the best documented example is the synergistic effect between radon progeny and cigarette smoking in producing lung cancer in underground miners. Pooled analysis of data that contained information from cohorts of underground miners on both exposure to radon progeny and smoking showed the presence of synergism. Documented examples of synergism between smoking and other health risk factors include asbestos and lung cancer (Mauderly and Samet, 2009).

In summary, selective Mauderly and Samet (2009) review confirmed that synergisms (greater than additive effects) among air pollutants in causing measurable biological effects have been demonstrated in laboratory studies of humans and animals. Report also identified evidence for additive and less than additive responses to combined exposures to multiple pollutants, sometimes among different outcome measures in studies in which certain outcomes revealed synergisms, and sometimes among the same outcomes at different times after exposure. Considering the likelihood that both synergisms and antagonisms result from environmental exposures and their importance to accurate assessments of risk, evaluations of pollutant interactions by both epidemiologic and laboratory research approaches will be very important for developing a stronger foundation for multipollutant air quality management (Mauderly and Samet, 2009).

Analyzing the types of environmental effects, we should clarify the difference between 'contamination' and 'pollution'. GESAMP (Joint Group of Experts on the Scientific Aspects of Marine Pollution) defines ***pollution*** as "the introduction by man, directly or indirectly, of substances or energy into the marine environment (including estuaries) resulting in such deleterious effects as harm to living resources, hazards to human health, hindrance to marine activities including fishing, impairment of quality for use of sea water and reduction of amenities". ***Contamination,*** on the other hand, is the "presence of elevated concentrations of substances in the environment above the natural background level for the area and for the organism" (FAO, 2006). Thus, dealing with Industrial ecology issues we should use the term 'pollution'.

The major forms of pollution are:

- Air Pollution
- Water Pollution
- Soil Pollution
- Radioactive Pollution
- Vibration
- Noise Pollution
- Thermal Pollution
- Landscape Pollution
- Effects on Biota
- Effects on Human Health

AIR POLLUTION

Air pollution is the emission of chemical substances and particles into the atmosphere. This kind of pollution results from the adulteration of pure air which exists in nature. There are many pollutants released in the air which dilute the natural/original quality. Common gaseous pollutants include carbon monoxide, sulfur dioxide, chlorofluorocarbons (CFCs) and nitrogen oxides. Most of these pollutants are emitted from industrial plants. Automobiles also play a major role in raising the levels of air pollution. Photochemical ozone and smog are created as nitrogen oxides and hydrocarbons enter the reaction with sunlight. It was at the onset of the Industrial Revolution that air pollution started affecting our world to a great extent.

It is difficult to classify air pollutants neatly because of the wide variety of matter emitted into the atmosphere by natural and anthropogenic sources. Atmospheric pollutants could be classified in different ways:

According to Origin:

- Primary air pollutants are directly emitted from the sources into the atmosphere. Typical air pollutants included under this category are:
- Inorganic gases such as SO_2, NO_X, H_2S, CO, NH_3, CO_2, HF;
- Particulate matter like ash, smoke, dust, fumes, mist, spray;
- Olefinic and aromatic hydrocarbons;
- Radioactive compounds (Shah, 2016).
- Secondary pollutants are those which are derived from the primary pollutants due to chemical or photochemical reactions in the atmosphere. Pollutants such as SO_2, NO_2, O_3, sulfate and nitrate salts, aldehydes and peroxyacetyl nitrate (PAN) are included in this category (Shah, 2016).

According to State of Substances:

- Gaseous pollutants;
- Particulate pollutants;
- Metallic component;
- Radioactive pollutant;
- Aerosols;
- Pesticides.

Let us analyze these air pollutants, their origin, and environmental impacts.

Sulfur oxides, SO_X: Combustion of sulfur containing materials produces SO_2 and $S0_3$. Global emission of SO_2 from natural sources like biological decay (H_2S), sea spray and volcanic activity is 90, 40 and 5 million tons per year while from anthropogenic sources such as coal, petroleum, smelting and sulfur production, it is 92, 26, 14 and 2 million tons per year respectively (Shah, 2016).

Sulfur dioxide (SO_2) is a colorless gas with a suffocating odor. Fossil fuel combustion is the reason of almost all anthropogenic (human-caused) sulfur emissions which are dangerous for environment and human health as they are corrosive to organic materials and irritate the eyes, nose and lungs. All fossil fuels contain sulfur, and it is released in the form of

SO_2 during fossil fuel combustion. Sulfur contents in fossil fuels range between 0.1% and 4% in oil, oil by-products and coal. Historically, the use of coal in domestic heating was a major source of sulfur dioxide emissions (at least in the UK), but it has declined substantially over time (Bright, 2008). In the developed countries, with the introduction of low sulfur fuels and switch from coal to gas as well as the increasing the energy efficiency, the sulfur dioxide emissions were declined significantly since 1970.

The main artificial sources of sulfur dioxide emissions are:

- Electric power generation
- Petroleum refining
- Other combustion
- Commercial and residential use
- Combustion for industry use
- Extraction and distribution of fossil fuels
- Automobile transport
- Aviation
- Water transport
- Railroad transport) (Paliulis, 2012).

It is considered today that the most significant sources of sulfur dioxide emissions are the electric power generating plants because of huge amounts of fossil fuel combustion; and the biggest sulfur dioxide emitters are considered to be China, Russia and India. Irina Bright (2008) admits that just one Siberian city in Russia – Norilsk – produces 1% of the total global emissions of sulfur dioxide. In 2007, Norilsk was announces to be one of the most polluted places on Earth (Bright, 2008).

But not all of sulfur dioxide emissions are caused by industrial activity. There are also significant sulfur emissions generated by natural sources (volcanoes, marine phytoplankton, vegetation etc.). The most natural sulfur emissions come in the reduced forms of sulfur compounds such as (Paliulis, 2012):

- hydrogen sulfide (H_2S);
- carbon disulfide (CS_2);
- carbonyl sulfide (COS);

and in the organic forms of:

- methyl mercaptan (CH_3SH);
- dimethyl sulfide (DMS), (CH_3SCH_3);
- dimethyl disulfide (DMDS), (CH_3SSCH_3);

As scientists admit, most of these compounds get oxidized to sulfur dioxide or to sulfate aerosols in the atmosphere. Most recently the natural sources have been by far surpassed by anthropogenic sources. Natural sources have been estimated to produce around 24% of all sulfur dioxide emissions, whereas human-caused emissions made up around 76% (Paliulis, 2012).

Marine phytoplankton produce dimethyl sulfide (DMS) which is then oxidized to SO_2 in the atmosphere; decay processes in soil and vegetation produce H_2S (as one of sulfur compounds); and SO_2 is emitted into the atmosphere by volcanoes. According to calculations, 90% of all natural sulfur emissions come in the form of DMS (Farmer, 1997). Sulfur dioxide, emitted in the natural air, produces such effects:

Acid rains.

Irritating eyes, nose, throat

Damage to lungs when inhaled

Acid rain, or acid deposition, is a broad term that includes any form of precipitation with acidic components, such as sulfuric or nitric acid that fall to the ground from the atmosphere in wet or dry forms. This can include rain, snow, fog, hail or even dust that is acidic (EPA, 2017). Figure 47 shows mechanism of acid rains formation.

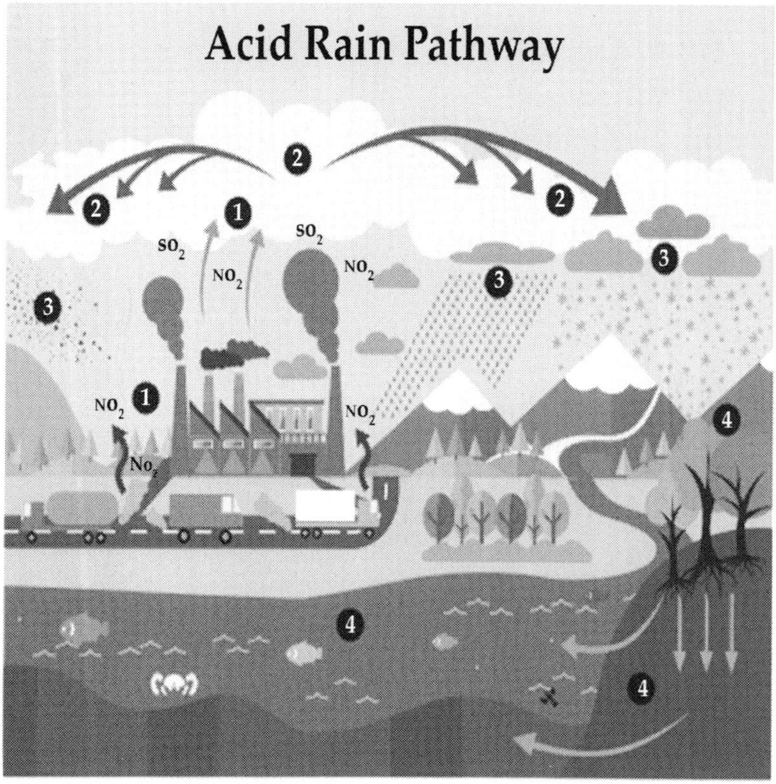

Figure 47: The Pathway of Acid Rain in the Environment.

Source: EPA (2017). https://www.epa.gov/acidrain/what-acid-rain

Acid rain results when sulfur dioxide are emitted into the atmosphere and transported by wind and air currents. The SO_2 reacts with water, oxygen, and other chemicals to form sulfuric acids. These then mix with water and other materials before falling to the ground. Acid precipitation leads to such environmental effects (Paliulis, 2012):

- rising the acidity of lakes and streams;
- damage to plant and fish life in lakes and streams;
- depletion of mineral nutrients in the soil;
- reduction of forest and agricultural yields;
- corrosion of metals;
- damage of buildings' surface.

Nitrogen Oxides, (NO_X): The oxides of nitrogen are N_2O, NO, NO_2, N_2O_3, N_2O_5. According to Sach (2016), the atmospheric background concentrations are: N_2O (0.25 ppm), NO_x (0.1 to 2 ppm) and NO_X (0.5 to 4 ppm). In air, NO_X is converted into nitric acid by natural processes which is one of the constituents of acid rain (Shah, 2016).

Andrew Farmer (1997) points out that oxides of nitrogen are produced by combustion of all fossil fuels including coal- and gas-fired power stations and motor vehicles. The fuel itself can produce some nitrogen (i.e., oil and coal contain around 0.5 – 1.5% of nitrogen, and natural gas contains less than that), but most of nitrogen oxides' production comes from the reaction of atmospheric nitrogen and oxygen within the combustion chamber (Farmer, 1997). The two main nitrogen oxides are nitrogen monoxide (NO) and nitrogen dioxide (NO_2). Nitrogen monoxide is a colorless gas, and nitrogen dioxide (NO_2) is a gas of reddish-brown color with a distinct sharp, biting odor. Combustion of fuels always produces both NO_2 and NO (Colls, 2002).

Around 90% of the NO_X pollutants comes from NO emissions which are then oxidized to nitrogen dioxide (NO_2) in the air. Therefore, only a small percentage of NO2 found in the atmosphere is directly emitted there in this formula. The rest has been formed because of chemical reactions in the atmosphere itself (Paliulis, 2012).

The most significant industrial sources of nitrogen oxides emissions are:

- Transport;
- Energy generation;
- Petroleum refining;
- Combustion for industry use;
- Production processes;
- Extraction and transportation of fossil fuels.

Automobile transport is considered to be the largest contributor of nitrogen emissions into the air. In the United Kingdom, vehicles contributed close to the half of all nitrogen emissions, followed by contributions from electric power generating plants which only contributed around 20% of total nitrogen emissions (Farmer, 1997). Environmental impact of nitrogen oxides emissions is presented at the Figure 48.

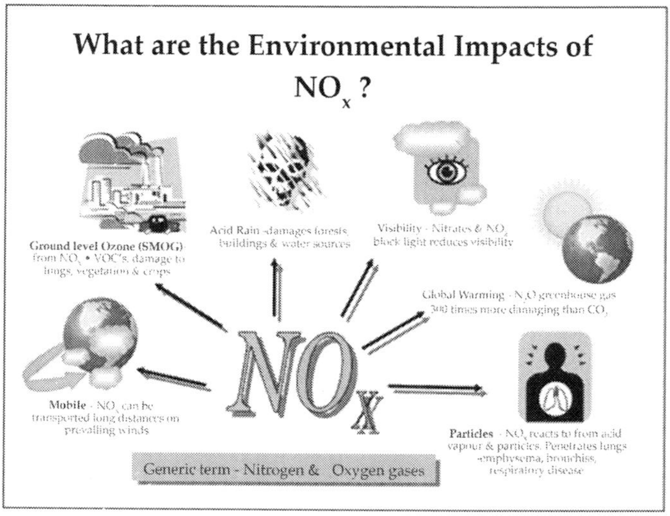

Figure 48: Environmental Effects of Nitrogen Oxides Emissions.

Source: Environmental Images. Studyblue. Available at: https://www.studyblue.com/notes/note/n/115-ways-to-go-apes/deck/16284387

Other environmental effects of nitrogen oxides emissions include (Bright, 2008):

- Pulmonary edema;
- Acid rain;
- Photochemical smog.

Carbon monoxide, CO: Carbon monoxide is a colorless, odorless gas which is highly toxic to humans (Shah, 2016). Carbon monoxide, produced by incomplete combustion of fuel, is the major individual pollutant with a tonnage matching that of all other pollutants together. Global emission of CO from natural sources amounts to 3364 million tons per year, while anthropogenic sources of CO emission, amounting to 95.7 MT/year, include:

- Transportation (70 MT/year),
- Fuel combustion (1.6 MT/year),
- Industrial processes, (7.8 MT/year),
- Solid waste disposal (7.8 MT/year),
- Others (8.5 MT/year) (Shah, 2016).

Motor vehicles and industry are considered to be the largest anthropogenic sources of carbon monoxide emissions which are fatal poisoning in many countries around the world. Dangerous effects of carbon monoxide emissions include:

- Problems with getting oxygen supplied to some body parts, especially to brain, which may be life-threatening (CO blocks hemoglobin in human blood from joining oxygen and taking it to the organs);
- Toxicity of the central nervous system and heart;
- Severe effects on the baby of a pregnant woman;
- Headaches and dizziness (Paliulis, 2012).

Ammonia: Ammonia is a colorless, pungent, hazardous caustic gas composed of nitrogen and hydrogen(Paliulis, 2012). Ammonia emissions are also grouped as NHy which is a sum of NH_3 and NH_4 (Colls, 2002).

According to Colls (2002), the main sources of ammonia emissions include:

- Livestock
- Fertilizer application
- Oceans
- Vegetation
- Biomass burning

Agriculture is the biggest contributor of ammonia emissions. Livestock farming and animal waste contribute the highest percentage of total ammonia emissions (due to the decomposition of urea from large animal wastes and uric acid from poultry wastes). Cattle accounts for more than 50% of all emissions (Paliulis, 2012). Exposure to very high levels of gaseous ammonia in the air may result in lung damage and even death (Bright, 2008).

Volatile organic compounds: Volatile organic compounds (VOCs) are defined as organic substances which easily evaporate and enter the atmosphere (Bright, 2008). VOCs include a wide range of air pollutants with organic origin, from pure hydrocarbons to incompletely oxidized hydrocarbons to organic compounds containing chlorine, sulfur, or nitrogen (Encyclopedia Britannica, 2008). Initially, the definition of VOCs did not include methane compounds (non-methane VOCs: NMVOCs) because it was considered that atmospheric concentration of methane has stable natural background. With the time scientists understood that methane is mostly an anthropogenic air pollutant that comes from fuel combustion, animal breeding, and rice production (Colls, 2002). According to Farmer (1997), though some of these compounds can have direct toxic effects, they have been grouped together because of their role in ozone formation.

Talking about natural emissions of volatile organic compounds, C. D. Geron and his colleagues (1994) admitted that forests are the primary natural sources of VOC emissions. According to Kesselmeier (2000), tropical forests are estimated to produce about half of all global natural non-methane VOC emissions.

Plants produce various organic molecules and release some VOCs (including a series of terpenes) into the atmosphere. In total, close to 1000 different compounds are known to be emitted by natural sources; and some of them are families with thousands of their own members (Colls, 2002).

The main industrial sources of volatile organic compounds include (Bright, 2008):

- Solvent use (including paints, adhesives, aerosols, metal cleaning and printing);
- Road transport (emissions from fuel / petroleum use);
- Production processes;
- Extraction and distribution of fossil fuels.

Jeremy Colls (2002) points out that substantial NMVOC emissions occur during the following processes:

- Painting (evaporation of thinners);
- Oil production (spreading and venting of gas);
- Oil refining (spreading and fugitive emissions);
- Transportation of oil or refinery products (evaporation from storage, displacement losses when venting tanks);
- Dry cleaning (final drying of clothes);
- Production of alcohol (breweries and distilleries);

- Arable farming (crop growing, silage manufacture, sludge spreading).

Emitted in the atmosphere, VOCs may produce the following effects (Bright, 2008):

- Some aromatic compounds such as benzene, toluene and xylene have potential carcinogenic effect and may cause leukemia and bladder cancer;
- As enablers in ozone formation, volatile organic compounds indirectly contribute to respiratory problems and other ozone-related problems;
- Contribute to sick building syndrome indoors.

Ozone (O_3): Ozone is a colorless, poisonous gas with a sharp, cold, irritating odor (Paliulis, 2012). Naturally, ozone can be found in the stratosphere, one of the upper layers of the atmosphere, where it occurs naturally; and in the troposphere, the lowest layer of the atmosphere, where it occurs both naturally and as a result of human-generated emissions (Colls, 2002).

The natural stratospheric ozone is beneficial for nature as it keeps harmful extreme ultraviolet sunlight from reaching the surface of the Earth. Ozone which is formed in the troposphere as a result of anthropogenic emissions of primary pollutants, has negative effects on human health and the natural environment (Bright, 2008). In that case ozone is considered to be an air pollutant.

Artificial-caused ozone in the troposphere is a secondary pollutant because it is produced by the reaction of primary pollutants, nitrogen oxides and hydrocarbons, including VOCs, in the presence of sunlight (Farmer, 1997). Farmer (1997) noticed that the process of ozone formation may take several days to complete, and ozone itself may turn out to be far from the sources of original primary pollutant emissions.

Ozone as an air pollutant in the troposphere can have the following negative effects on the natural environment, biodiversity, and human health:

- Irritation of the respiratory system that results in coughing, throat sore and an uncomfortable sensation in the chest;
- Vulnerability to respiratory infections;

- Compromised lung function damaging the breathing process that may become more rapid and more shallow than normal;
- Inflammation and damage to the lungs lining;
- Aggravation of asthma;
- Reduction in agricultural yields;
- Interference with photosynthesis and suppression of growth of some plant species (Bright, 2008).

Tropospheric ozone is the main cause of the photochemical smog. Photochemical smog has brown-yellow color; it is a product of the chemical reaction between sunlight, nitrogen oxides and VOCs, which results in the formation of ozone and airborne particles (Paliulis, 2012). Both industrial smog and photochemical smog are forms of atmospheric pollution. Although we can see both as yellow-brown haze, they differ in two key ways: chemistry and composition. According to Harvey (2017), industrial smog typically exists in urban areas where factories burn fossil fuels such as coal, which creates smoke and sulfur dioxide that mix with fog droplets to create a thick blanket of haze close to the ground. Photochemical smog occurs in drier, sunny areas and occurs as a result of increased usage of all fossil fuels, including gasoline, and the burning of trees and organic waste. These emissions mix and react in the presence of sunlight and create a noxious environment that can be harmful to human health (Harvey, 2017).

According to the U.S. Environmental Protection Agency, "photochemical smog is a mixture of pollutants that form when nitrogen oxides and volatile organic compounds react to sunlight, creating a haze that is common over cities" (EPA, 1995). The two major primary pollutants, nitrogen oxides and volatile organic compounds, combine in a range of chemical reactions to create secondary pollutants, which are dangerous when detected in our atmosphere and at ground level as it is shown at the Figure 49. The couple of dominant toxic components created in photochemical smog are ozone and peroxyacetyl nitrate. While ozone can be produced naturally in an unpolluted environment, the interaction with VOCs prevents nitrogen oxide – after being transformed to nitrogen dioxide – from consuming ozone. This process leads to toxic, harmful levels of ozone in the immediate environment (Harvey, 2017).

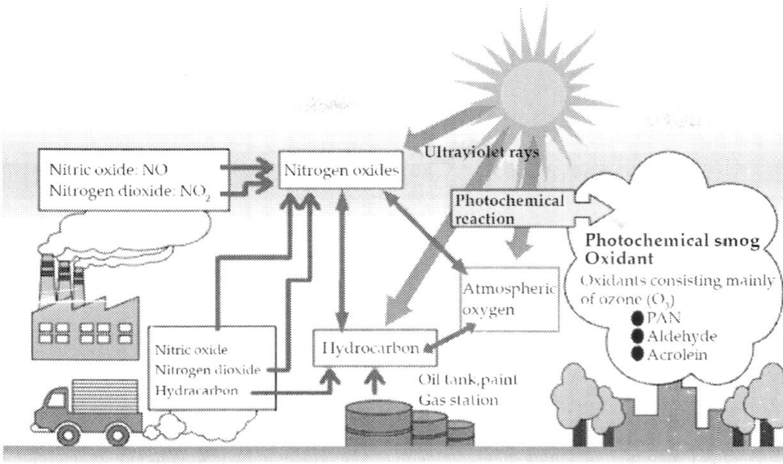

Figure 49: Mechanism of Photochemical Smog Formation.

Source: Photochemical Smog. http://www.markedbyteachers.com

As Samanta Harvey (2017) informs, industrial smog contains two primary components: sulfur dioxide and particulates, which include dust and soot from combusting coal for heat and fuel. Combined with the water, the mixture turns into fog to produce acid rain. The sulfur dioxide, produced during coal or other fossil fuels burning, dissolves in water and becomes a toxic stew of thick fog and particles. Firstly, industrial smog was detected at the time of Industrial Revolution and was most prevalent in the past 50 years with the heating of coal for heat and fuel (Harvey, 2017).

According to the National Institute of Health, both types of smog can cause major health risks, including asthma, lung tissue damage, bronchial infections, and heart problems. Toxic concentrations of ozone created by photochemical smog can damage lung tissue and destroy the immune system. The sulfuric acid created by industrial smog is responsible for thousands of fatalities beginning with the Industrial Revolution and continuing until today. Industrial and photochemical smog can also cause breathing problems and skin irritations, and can even corrode buildings (Harvey, 2017).

Persistent organic pollutants: Persistent organic pollutants (POPs) are substances which are resistant to degradation and persistent in the environment, with half lives of years in the soil or sediment and days in the atmosphere (Paliulis, 2012). This group of contaminants includes dioxins, furans, polychlorinated biphenyls (PCBs) and organochlorine pesticides such as DDT (Colls, 2002).

These compounds enter the food chains via the process of biomagnification, get accumulated in human and animal tissue, and are capable of long range transport through being attached to airborne particles (Paliulis, 2012).

The sources of persistent organic pollutants have the artificial origin. In agriculture are used as pesticides. DDT was broadly used as a pesticide in the middle of twentieth century, and it is still can be found in the organisms of penguins at Antarctic. Other POPs are used in industrial manufacturing as well as in the production of goods such as solvents, polyvinyl chloride, and medicines (Paliulis, 2012).

Exposure to persistent organic pollutants takes place through diet, especially, through the consumption of animal fats, environmental exposure or accidents. POPs may lead to (Bright, 2008):

- Death and illness including disruption of endocrine, reproductive and immune systems;
- Neurobehavioral disorders;
- Cancers.

When persistent organic pollutants are present in the atmosphere in the form of aerosols, they may be classified as airborne particles rather than gaseous pollutants (Paliulis, 2012).

Particulate Pollutants / Airborne Particles. According to Shah (2016), small solid particles and liquid droplets are termed particulates. Their number varies from several hundred per cm^3 in clean air to more than 100,000 per cm^3 in highly polluted air. Every year natural sources discharge 800 to 2000 million tons of particulates. Industrial sources are responsible for 200 to 500 million tons of particulate pollutants annually (Shah, 2016).

Particulate pollutants are tiny fragments of solid or liquid nature suspended in the air (aerosols). Particles may be primary (when they are emitted directly into the atmosphere by sources), or secondary (when they are formed in the atmosphere through the interaction of primary emissions). Solid particles of sizes between 1 and 100 μm (micrometers) in diameter are called dust particles, and solid particles less than 1 μm in diameter are called fumes, or smoke (Encyclopedia Britannica, 2008).

Natural sources of particulates are (Bright, 2008):

- Erosion of soil by wind which generates dust particles that travel around the globe;
- Evaporation of droplets of sea water resulting in sea salt crystals being suspended in the air;
- Volcanoes;
- Forest fires;
- Living vegetation.

Industrial sources of particulates are:

- Fossil fuel combustion (fly ash from power plants);
- Road transport;
- Power generating plants;
- Production processes (such as dust blown away by winds from construction sites, emissions from smelters, refuse burning etc.) (Paliulis, 2012).

According to Irina Bright (2208), airborne particles may cause:

- Asthma;
- Lung cancer;
- Cardiovascular problems.

Particles less than 10 μm in diameter can be easily inhaled and get trapped in the respiratory system, so they are the biggest concern to human and animal health (Paliulis, 2012). Jeremy Colls (2002) also notes that particles of this size have very low gravity-related sedimentation rates and may therefore remain in the atmosphere for days before being washed out by rain or attached to vegetation or buildings.

Greenhouse Gases (GHG): Gases that trap heat in the atmosphere are called greenhouse gases. US Environmental Protection Agency pointed out such greenhouse gases (EPA, 2017):

Carbon dioxide (CO_2): Carbon dioxide enters the atmosphere through burning fossil fuels (coal, natural gas, and oil), solid waste, trees and wood products, and also as a result of certain chemical reactions (e.g., manufacture of cement). Carbon dioxide is removed from the atmosphere (or "sequestered") when it is absorbed by plants as part of the biological carbon cycle.

Methane (CH_4): Methane is emitted during the production and transport of coal, natural gas, and oil. Methane emissions also result from livestock and other agricultural practices and by the decay of organic waste in municipal solid waste landfills.

Nitrous oxide (N_2O): Nitrous oxide is emitted during agricultural and industrial activities, as well as during combustion of fossil fuels and solid waste.

Fluorinated gases: Hydrofluorocarbons, perfluorocarbons, sulfur hexafluoride, and nitrogen trifluoride are synthetic, powerful greenhouse gases that are emitted from a variety of industrial processes. Fluorinated gases are sometimes used as substitutes for stratospheric ozone-depleting substances (e.g., chlorofluorocarbons, hydrochlorofluorocarbons, and halons). These gases are typically emitted in smaller quantities, but because they are potent greenhouse gases, they are sometimes referred to as High Global Warming Potential gases (High GWP gases).

Greenhouse gases are the main factor in developing the greenhouse effect. The Greenhouse Effect is a natural process that warms the Earth, and, in fact, is quite necessary for our survival. They act as a natural blanket by preventing the sun's heat energy from radiating back into space, much like a greenhouse traps the sun's energy to warm someone's plants even in the middle of winter. The natural greenhouse effect helps warm the Earth's surface by as much as 33°C, and without it, our planet would be too cold for humans to survive (State Delaware, 2013).

Figure 50 illustrates the basic processes that form the greenhouse effect. As the sun's energy hits the Earth, some of that energy is absorbed by the earth's crust and by the oceans, warming the planet. The rest of the energy is radiated back toward space as infrared energy. While some of this infrared energy does radiate back into space, some portion is absorbed and re-emitted by the layer of greenhouse gases in the atmosphere. This absorbed energy helps to warm the planet's surface and atmosphere just like a greenhouse.

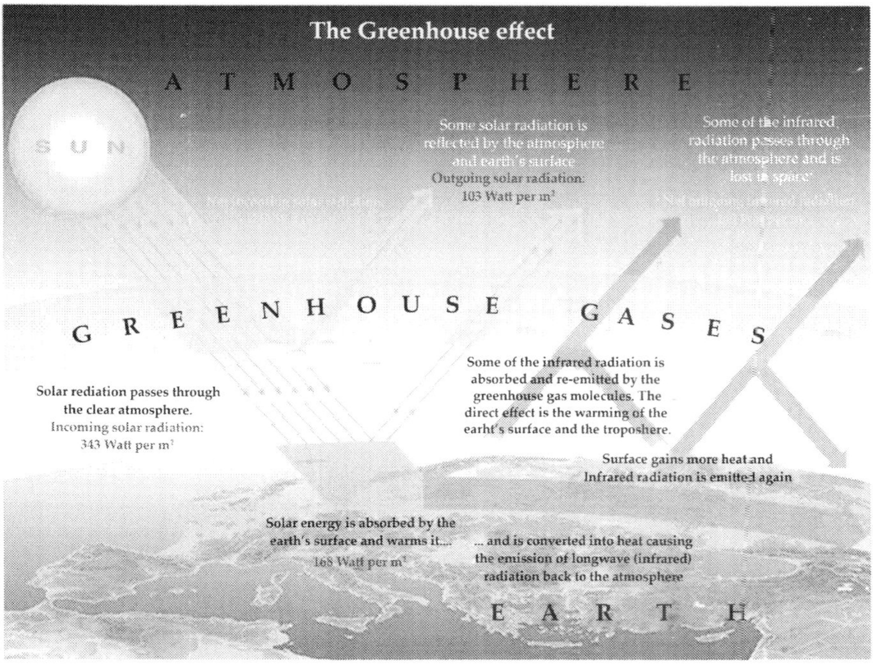

Figure 50: The Mechanism of Greenhouse Effect.

Source: GRID, UNEP and WMO, 1998.

There are three main factors that directly influence the greenhouse effect: (1) the total energy influx from the sun, which depends on the earth's distance from the sun and on solar activity, (2) the chemical composition of the atmosphere (what gases are present and in what concentrations), and (3) albedo, the ability of the earth's surface to reflect light back into space. The only factor that has changed significantly in the last 100 years is the chemical composition of the atmosphere—and that is because of human activity (EPA, 2017).

Although the greenhouse effect is a naturally-occurring process, humans have recently amplified the natural effect by increasing the concentration of CO_2, water vapor, and other greenhouse gases in the atmosphere (primarily through fossil fuel combustion in the energy and transportation sectors).

According to the US EPA, Earth's average temperature has risen by 1.5°F over the past century, and is projected to rise another 0.5 to 8.6°F over the next hundred years. Small changes in the average temperature of the planet can translate to large and potentially dangerous shifts in climate and weather. The evidence is clear. Rising global temperatures have been accompanied by changes in weather and climate. Many places have seen changes in rainfall, resulting in more floods, droughts, or intense rain, as well as more frequent and severe heat waves. The planet's oceans and glaciers have also experienced some big changes – oceans are warming and becoming more acidic, ice caps are melting, and sea levels are rising. As these and other changes become more pronounced in the coming decades, they will likely present challenges to our society and our environment (EPA, 2017).

WATER POLLUTION

Water pollution can be defined in many ways. Usually, it means one or more substances have built up in water to such an extent that they cause problems for animals or people. Water pollution almost always means that some damage has been done to an ocean, river, lake, or other water source. A 1969 United Nations report defined ocean pollution as: "the introduction by man, directly or indirectly, of substances or energy into the marine environment (including estuaries) resulting in such deleterious effects as harm to living resources, hazards to human health, hindrance to marine activities, including fishing, impairment of quality for use of sea water and reduction of amenities" (UN, 1969).

Water pollution happens when toxic substances enter water bodies such as lakes, rivers, oceans and so on, getting dissolved in them, lying suspended in the water, or depositing on the bed and degrading the quality of water. Water pollution is caused by the discharge of wastewater from commercial and industrial waste (intentionally or through spills) into waters; discharges of untreated domestic sewage, and chemical contaminants into waters (including urban runoff and agricultural runoff, which may contain chemical fertilizers and pesticides).

Food and Agriculture organization of United Nations provide such classification of water pollutants (FAO, 1999):

Suspended solids: According to the UN Food and Agriculture Organization (2006), presence of suspended solids in water gives rise to turbidity. Suspended solids may be the components of clay, silt, airborne particulates, colloidal organic particles, plankton, and other microscopic organisms. The presence of particulates in water, whether organic, inorganic, or micro-organisms origin, can protect bacteria and viruses from the action of disinfectants (FAO, 2006). The adsorptive ability of some suspended particulates can lead to entrapment of undesirable inorganic and organic compounds present in the water and in this way, turbidity can bear an indirect connection to the health aspects of water quality.

Airborne particulate matter is important concern to facilities located near mineral stockpiles (coal, iron ore, bauxite, etc.) or downwind from large power stations (fly ash), timber saw mills (saw dust) or cement factories (cement dust). Precipitation water collection systems are particularly sensitive to such airborne particulate matter because it usually augments local potable water systems and acts as conduits for pollutants to enter potable water systems. Large amounts of aromatic hydrocarbons are also generated by the fossil fuel combustion in oil-fired power stations and industrial kilns (FAO, 2006).

Hardness: Water hardness is the traditional ration of the ability of water to react with soap, hard water requiring a considerable amount of soap to produce a lather (FAO, 2006). The level of water hardness has been classified in terms of its equivalent $CaCO_3$ concentration as follows:

- soft – 0 to 60 ppm;
- medium hard – 60 to 120 ppm;
- hard – 120 to 180 ppm;
- very hard – 180 to and above (FAO, 2006).

According to FOA (2006), soft water has a greater tendency to cause corrosion of pipes and consequently, certain heavy metals such as copper, zinc, lead and cadmium may be present in the water. Very hard water, on the other hand, can cause considerable incrustations in pipes and fittings, especially in fish processing plants.

pH: The value of the pH, as it is defined in FAO (2006) document, expressed as a value ranging between 1 and 10, is a good indicator of the state of the water. Values of 9.5 and above are alkaline in taste. Values of 3 and below are acidic in taste. Values lower than 6 cause problems with corrosion. Values below 4 support little life in a marine environment. Drinking water should have a pH in the range 6.5 to 8.5. Harbor water should have a pH of between 6 and 9 (FAO, 2006).

Sodium: The sodium ion is universal in water leading to the high solubility of its salts and the abundance of mineral deposits. Near coastal areas, sea breeze can make an important contribution to sodium water pollution, either by fallout on to land surfaces where it drains to the water source or from washout by rain. Domestic, commercial, and industrial sewage waters are another source of sodium in water. In their nature, sodium salts are not acutely toxic substances kidneys are able to excrete sodium, but effects on infants, in opposite to adults, are different because of the immaturity of infant kidneys (FAO, 1999).

Sulfates: UN Food and Agriculture Organization (2006) informs that sulfates are widely distributed in nature and excessive amounts of dissolved sulfates in drinking water lead to problems with hardness.

Chlorides: According to the UN Food and Agriculture Organization (1999), chlorides are widely distributed in nature and are present in mineral deposits, in sea water and some industrial processes.

Hydrogen Sulfide: Hydrogen sulfide is present as a by-product in septic tanks when proteins in sewage are attacked by anaerobic kinds of bacteria (FAO, 1999).

Nitrates and nitrites: Nitrates and nitrites are considered by the UN Food and Agriculture Organization (2006) together because conversion from one form to another occurs in the environment. The health impacts of nitrate are generally a result of its ready conversion to nitrite in the body. Nitrates are products of oxidation of organic nitrogen by the bacteria present in soils and in water where sufficient oxygen is present. Incomplete bacterial oxidation of organic nitrogen forms nitrites. One of the principal way to nitrate usage is as fertilizer. Nitrates are also used in explosives, as oxidizing agents, and in the chemical industry, and also as food preservatives. their occurrence in water is brought about by indiscriminate fertilizer use, decomposing vegetable and animal matter, sewage effluents, industrial discharges and last but not least, leachates from refuse dumps (FAO, 1999).

Petroleum Hydrocarbons: Human's taste threshold for petroleum hydrocarbons is very low; their taste is particularly repulsive. All components of crude oil are degradable by bacteria, though at varying rates, and a variety of fungi and yeasts can also process petroleum hydrocarbons. Water-soluble components of crude oils and refined products include a variety of compounds that are toxic. High-molecular-weight tars are less harmful in the water than medium-molecular-weight components such as diesel. Low-molecular-weight components are generally less important because they are very volatile and rapidly evaporate. Polluted harbor water can impart an unfriendly flavor to fish if used for cleaning. Commercial fishermen may also risk tainting a whole catch if their nets have been fouled by oil or diesel pollution (FAO, 2006).

Biodegradable organics: Composed primarily of proteins, carbohydrates, and fats, biodegradable organic is measured most commonly in terms of Biological Oxygen Demand (BOD). Biological Oxygen Demand is the quantity of oxygen required for the oxidation of organic matter by bacterial action in the presence of oxygen. The higher the consumption of oxygen (the more organic the pollution) the less oxygen is left to support life (FAO, 2006).

Aluminum: According to the UN Food and Agriculture Organization (2006), aluminum does not consider to be an essential nutrient to man. Compared with the aluminum intake from food, aluminum that comes into human body with water pollution is small.

Copper: Copper increases corrosion of aluminum and zinc fittings, stains clothes and plumbing fixtures. This metal is used in alloys, as a catalyst, in anti-fouling paints and as a wood protector. Urban sewage contains significant amounts of copper (FAO, 2006).

Iron: The presence of iron and its compositions in water is objectionable for many reasons unrelated to health. Under the pH conditions existing in water bodies, ferrous salts are unstable and precipitate as insoluble ferric hydroxide, which looks as rusty silt. The taste of that water is unpalatable, it promotes the growth of 'iron bacteria' and the silt gradually decreases the flow of water in the piping (FAO, 1999).

Manganese: UN Food and Agriculture Organization (2006) informs that anaerobic groundwater often contains elevated levels of dissolved manganese. that imparts an undesirable taste to beverages and stains plumbing fixtures.

Zinc: The concentration of zinc in tap water can be significantly higher than that in surface water leading to the leaching action of zinc from galvanized pipes, brass and other zinc alloys. Zinc makes water to taste undesirable, even in concentrations of 5 ppm. The water may appear opalescent and grow a greasy film during boiling (FAO, 1999).

Priority pollutants: Priority pollutants are organic or inorganic substances selected on the basis of their detected or suspected carcinogenicity, mutagenicity, teratogenicity or high acute toxicity (FAO, 1999).

Arsenic: Arsenic is notorious by the UN Food and Agriculture Organization (1999) as a toxic element. Its toxicity, however, rest on the chemical (valence) and physical form of the component, the way by which it enters the body, the dose and duration of exposure and several other biological parameters. It is recommended that, when water is found to be polluted by arsenic at levels of 0.05 ppm, an attempt should be made to determine the valence and chemical forms of the element. Arsenic is used in industrial production as an alloying additive with lead solder, lead shot, battery grids, cable sheaths and boiler piping. Nowadays, most arsenic originates from paints or pharmaceuticals and is commonly found in sewage. The average concentration of arsenic in sea water is around 0.002 ppm (FAO, 1999). Health effects of this metal include carcinogenicity and mutagenicity.

Asbestos: According to the UN Food and Agriculture Organization (1999) report definition, asbestos is a general term for fibrous silicate minerals of the amphibole and serpentine mineral groups. There are 6 minerals that are characterized as asbestos: chrysotile, crocidolite, anthophyllite, tremolite, actinolite, and amosite. Asbestos is often found in domestic water supplies because of the use of asbestos cement (170 g of asbestos per kg – 80% chrysotile and 20% crocidolite) for pipes in distribution systems (FAO, 2006). Asbestos is proved as highly carcinogenic material.

Barium: UN Food and Agriculture Organization (2006) informs that barium is used in various industrial processes, such as in vacuum tubes, spark-plug alloys, Getter alloys, Fray's metal and as a lubricant for anode rotors in X-ray machines. It is recommended by FAO that drinking water should not contain more 0.050 ppm.

Beryllium: Beryllium is commonly found as part of feldspar mineral deposits and can exist as the mineral beryl in small localized areas. The main source of beryllium in the environment is the combustion of fossil fuels, although contamination is normally light. Beryllium enters the water system through weathering of rocks in ground aquifers, atmospheric fallout on rain water collection systems and with industrial and municipal sewage waters. In industrial production beryllium is used in metal alloys and certain electrical components (FAO, 2006).

*Selenium:*_UN FAO informs that selenium is widely used in the electronics industry, TV cameras, solar batteries, computer cores, rectifiers, xerographic plates, and in the ceramics as a colorant for glass. It is also used as a trace element for animal feeding. There is a range of selenium intake by humans that is consistent with health, but outside of safety limits deficiency or toxicity can occur (FAO, 1999).

Silver: Silver can be found naturally in elemental form and as various ores. It is usually associated with lead, copper, and zinc ores. Silver levels in tap water may sometimes be elevated because such metals as lead and zinc are used in distribution systems and also because in some countries silver oxide is used to disinfect water supplies. The levels of silver in drinking water should not exceed 1 ppb. In manufacturing process, silver is used to produce the silver nitrate, silver bromide and other photographic chemicals, for water distillation equipment, mirrors, silver plating equipment, special batteries, table cutlery, jewelry, dental medical and scientific equipment including amalgams (FAO, 2006).

Dissolved inorganics: Dissolved inorganic contaminants are generally associated with the aesthetic and organoleptic (taste and odor) characteristics of water (FAO, 1999). Organoleptic characteristics of water are usually the concern of social, aesthetic, economic, and cultural considerations, but dissolved inorganic pollution also is a reason for the health risk.

Refractory organics: Refractory organic presents a wide range of pollutants including chlorinated alkanes (carbon tetrachloride), chlorinated ethenes (polyvinyl chloride or PVC), polynuclear aromatic hydrocarbons (naphtalene, coal tar), pesticides, herbicides and fumigants (DDT, Endrin, Aldrin, Lindane, Methoxychlor, Toxaphene and Silvex), Mono-Dichlorobenzenes (solvents), Benzenes (Benzene, Toluene),

Phenols and Chlorophenols and Trihalomethanes (Chloroform, Bromoform) (FAO, 1999).

Chlorinated Alkanes: Chlorinated alkanes are usually used in the chemical industry as an intermediate for the manufacturing of other organochlorine compounds. They are produced in large amounts and therefore many of them are found in raw and finished drinking water. For example, Carbon Tetrachloride (haloalkane with a wide range of industrial and chemical applications) usually is found during the disinfection of drinking water but is not produced in drinking water as a result of the chlorination process itself. During the previous time, it was broadly used as a propellant in aerosols production. This chemical can cause cancer (FAO, 2006).

Chlorinated Ethenes: Chlorinated ethenes are widely used in a assortment of industrial processes as solvents, softeners, paint thinners, dry-cleaning fluids, intermediates, etc. They have high volatility, therefore they are usually enter the atmosphere from surface water and their concentrations in water are usually lower. As an example, vinyl chloride is broadly used to produce PVC resins which, in turn, is used in production of most widely variety of plastics in the world. Vinyl chloride can cause cancer and is mutations in a number of biological systems (FAO, 1999).

Polynuclear Aromatic Hydrocarbons (PAH): PAH are a large group of organic compounds present in the environment from both natural and industrial sources. polynuclear aromatic hydrocarbons are rarely encountered singly in the environment and many interactions can occur with mixtures of polynuclear aromatic hydrocarbons whereby the potency of known carcinogenic PAH may be enhanced. Reaction with coal-tar based linings during distribution is believed in some instances to lead to an increase of PAH concentration in water. Aromatic hydrocarbons can enter the natural waters from discharges from ships as ballast water, bilge pumping, vehicle engines exhausts, effluents from coastal refineries, crude oil power stations, terrestrial run-off (particularly from urban storm water containing road asphalt particles) and leaching (creosoted components from jetties and wharves) (FAO, 2006).

Pesticides: Pesticides influence the water quality very much. This group includes chlorinated hydrocarbons and their derivatives, persistent herbicides, soil insecticides, pesticides that are easily leached out from the soil, and pesticides systematically added to water supplies to control disease vectors, such as mosquito larvae. For example, chlorinated hydrocarbon insecticides occur frequently on our fields and they are very persistent in the environment where they have become ubiquitous (FAO, 2006).

Mono-Dichlorobenzenes: UN Food and Agriculture Organization (2006) admits that mono chlorobenzenes are widely used as solvents and during the manufacturing of several chemicals (i.e., insecticides and phenols) while dichlorobenzenes are important intermediates for dyes, moth repellants, deodorants, dielectric fluids, heat transfer fluids and insecticides.

Benzenes: Benzene and Toluene are the result of petroleum manufacturing and using, or they can be the by-product in the manufacturing of gas. Both substances are widely used in the chemical industry as intermediates and to produce styrene, phenol, acetone, and cyclohexene (used in manufacturing nylon). Manufacturing of plastics, paints, detergents and petrol additives needs significant quantities of toluene (FAO, 2006).

Phenols and chlorophenols: According to the UN Food and Agriculture Organization (2006), chlorophenols are used as biocides and are found in water as a result of chlorinating water supplies containing phenol. Chlorophenols are well known for their low taste and odor thresholds (FAO, 1999).

Heavy metals: UN Food and Agriculture Organization (2006) argues that trace quantities of many metals are important constituents of most waters. But in amounts that are higher than natural level, heavy metals are also classified as pollutants. The presence of any of these metals in excessive quantities will restrict with many beneficial uses of the water because of their toxicity (FAO, 1999).

Cadmium: Cadmium is widely distributed in the Earth's crust, usually it is associated with zinc and copper (FAO, 2006). Commercially cadmium is produced as a by-product of zinc smelting, metal mining, smelting and fossil fuel combustion. Cadmium is highly toxic to the human body. UN FAO (1999) informs that cadmium and its compounds are black-listed materials, which by international agreement may not be discharged or dumped into the environment because of cadmium's cumulative poison effect.

Chromium: Chromium is very toxic in its soluble forms which are produce in manufacturing processes of chrome alloys, chrome plating, oxidizing agents, corrosion inhibitors, in leather industry, producing pigments for the textile, glass, and ceramic industries, as well as in photography. Hexavalent soluble chromium compounds are highly carcinogenic (FAO, 2006).

Lead: Lead is the most abundant of heavy metals occurring in nature. This heavy metal was one of the first metals used on a large scale by humanity. Lead is very dangerous to human health as a cumulative poison, so its monitoring is vital. Today the major amount of lead produced in metallic form, batteries, cable sheathing, sheets, and pipes, etc., is recovered and recycled. But some lead is used in compound form, like paints and petrol additives, is lost in the environment, finally ending up in the aquatic environment. Lead materials, similar to the ones used as petrol additives, are broadly used in the production of mercurial fungicides (FAO, 2006).

Mercury. Mercury is a comparatively rare element, but it is ubiquitous in the environment because of natural geological activity and man-made pollution. Mercury is extremely useful to man, but it is also highly toxic to the human body, especially in the form of methyl mercury, because it cannot be excreted, so it acts as a cumulative poison. The most famous example of lead harmful effect on human health is the Minamata disease in Japan that even caused mutation in next generations. UN FAO informs that potential for long-term human health hazards from ingesting mercury-contaminated fish has led several nations to establish regulations and guidelines for allowable sea-food mercury levels. Nearly all levels higher than 1 ppb in water are due to industrial effluents connected with chlorine and caustic soda production, pharmaceuticals, mirror coatings, mercury lamps and certain fungicides (FAO, 1999).

Nickel: Nickel is universal in the environment. Nickel is a relatively non-toxic element; however, according to the UN Food and Agriculture Organization (2006) report, some nickel materials have shown to be carcinogenic in animal experiments.

Tin: Tin and its compounds are significant but provocative chemicals in the environment. Similarly to other elements, not all chemical containig tin are equally biologically active. Inorganic tin (derived from eating canned foods) has low toxisity. In contrast, some organic tin compounds (organo-tins like Tributyltin and Triphenyltin) are highly toxic. In many countries, organotin anti-fouling paints are prohibited on vessels less than 25 meters long, and the start of a fishing season generally sees an increase of this compound in the water as freshly painted vessels are launched back into the water (FAO, 1999).

SOIL POLLUTION

Soil pollution is defined as "the presence of toxic chemicals (pollutants or contaminants) in soil, in high enough concentrations to pose a risk to human health and/or the ecosystem" (Environmental Pollution Centers, 2017). In the case of contaminants which present naturally in soil, even when their levels are not high enough to pose a risk, soil pollution is still said to occur if the levels of the contaminants in soil exceed the levels that should naturally be present (Environmental Pollution Centers, 2017).

The pollution of soil affects the entire ecosystem as it is the source of nutrition and food for all the living beings on earth. Any alteration of the purity of this natural resource has the effect of a 'chain-reaction'. Pollutants, contained in the soil, enter the food chain and disseminate to every nook and corner of the planet. Therefore, amongst the air, water, and soil types of environmental pollution, soil pollution is, the most dangerous. Apart from insecticides, pollutants in the soil are also emitted as a result of leaching of landfill wastes. The most common chemicals that are recognized as soil pollutants are: petroleum hydrocarbons, polynuclear aromatic hydrocarbons (such as naphthalene and benzo(a) pyrene), solvents, pesticides, lead, and other heavy metals (Shayler et.al., 2009). Soil pollution of certain area is usually correlated with the degree of industrialization and intensity of chemical usage of region.

Human activities polluting the soil range from industrial manufacturing, urban or industrial wastes, radioactive emissions that contaminate the soil with various toxic substances, to agricultural activities that infest the crops with pesticide chemicals. Environmental Pollution Centers pointed such kinds of soil pollutants (2017):

- Biological agents. Biological agents usually enter the soil from the human, bird, and animal excreta and work inside the soil to introduce manures and digested sludge;

- Agricultural pollutants. This group includes pesticides, fertilizers, herbicides, slurry, debris, and manure that are results of agricultural activity;
- Radioactive pollutants. Radioactive materials such as Radium, Thorium, Uranium, Nitrogen, etc. are emitted during industrial activity. They can infiltrate the soil and create toxic effects.
- Urban waste. Urban waste consists of garbage and rubbish materials, dried sludge, and sewage from domestic and commercial waste.
- Industrial waste. Steel, pesticides, textiles, drugs, glass, cement, petroleum, etc. are produced by paper mills, oil refineries, sugar factories, petroleum industries and others industries (Environmental Pollution Centers, 2017).

Main sources of soil pollution, according to Cornell Waste Management Institute (2009), are:

- Industrial / Commercial site use: Different chemicals that may be present due to industrial or commercial activities are depend on the type of industry and the specific procedures used on site. The level of contamination depends on many factors, such as how close to the property a activity occurred, and what is the duration of use (Shayler et.al., 2009).
- High traffic areas: Lead compounds were used in gasoline until the late 1970s; after this time, their use was phased out. Even though the use of leaded petrol has now been prohibited, the highest concentrations of lead in soils are still generally found adjacent to busy roadways. Polyaromatic hydrocarbons (PAHs) are chemicals associated with the incomplete burning of fossil fuels and with coal tars and asphalt. The levels of polyaromatic hydrocarbons and some other chemicals may also be higher in high traffic areas as compared to other areas (Shayler et.al., 2009).
- Treated lumber: Arsenic, in the form of chromated copper arsenate (CCA), was used in wood protectors to make pressure-treated lumber. CCA-treated lumber is no longer available in the US for residential uses, but it can still be used in industrial buildings. Some of the arsenic in CCA-treated wood can enter

from the wood to nearby soil, although it does not travel far from the wood structure (Shayler et.al., 2009).

- Petroleum spills: Leakage from gas stations, fuel tanks, or other activities can result in petroleum spills and elevated levels of pollutants in the soil such as benzene, toluene, and xylene in the soil. Some of these chemicals are volatiles, so it is prohibited to remain them in the surface soil where they would be taken up by plants or be in direct contact with humans, unless the spill was very recent or large. In some cases underground spills can happen, and they may result in vapors that make their way to the surface soil (Shayler et.al., 2009).
- Automobile or machine repair / junk vehicle storage: Automobile or machine repair services may result in accidental spills or prearranged dumping of pollutants into the soils. Such type of activity is associated with many possible chemicals and wastes, including petroleum products, PAHs (particularly from motor oil), solvents like trichloroethylene, used tires and rubber products, heavy metals (used engine oil may contain chromium, lead, molybdenum, or nickel from engine wear), or used batteries (which may release lead or mercury). Junk vehicles may also be a source of these or other pollutants, depending on their condition and how and where they are stored (Shayler et.al., 2009).
- Furniture refinishing: Some chemical strippers used in furniture refinishing have methylene chloride and other solvents, including toluene and methanol, inside. These substances can pollute the soil and even groundwater if handled improperly during manufacturing operations or projects by a home hobbyist (Shayler et.al., 2009).
- Landfills / Garbage dumps: Many different soil pollutants can leach from landfills or other garbage disposal sites, including petroleum products, solvents, pesticides, mercury, lead and other heavy metals. Such chemicals that may be present in soils nearby areas used for waste disposal (currently or in the past), depend on the specific conditions of a territory, and on garbage or by-products types that were disposed of at that site (Shayler et.al., 2009).

- Fires: The intentional or accidental burning of wastes or some materials can produce and release PAHs, dioxins, or other chemicals into soils, depending on what was burned and how long ago. Burning yard wastes, such as tree branches, is much less likely to discharge harmful pollutants than intentional or accidental fires that burn garbage, buildings and their inside furniture, or other synthetic substances (Shayler et.al., 2009).
- Fertilizers: Some fertilizers are produced from waste materials, i.e., sewage biosolids or fly ash, and their usage may result in the soil pollution with heavy metals (such as copper, zinc, cadmium, and lead) and PBTs (persistent, bioaccumulative, toxic chemicals) to soils. Fertilizers manufactured from cement kiln dust can also contain heavy metals and dioxins. Phosphate fertilizers manufactured from the rock phosphate are reported to contain some cadmium, and manures are sometimes relatively high in copper or zinc (Shayler et.al., 2009).
- Lead paint: Some paints produced before 1978 are likely to have lead inside. As lead paint getting older and peels off or is removed through stripping, scraping or sandblasting activities, lead can enter the soil surrounding homes or other buildings. Reported concentrations of lead in soils are usually highest right near a building, and tend to decrease with distance away from the contamination source (Shayler et.al., 2009).
- Pesticides: Pesticides group includes chemicals used as insecticides, herbicides, fungicides, rodent poisons and some other kinds of toxic substances. The best way to test the soil for pesticides is to consider if and how pesticides might have been used on the site, and to try to get information on what might have been used and where. Scientists report that chlordane, a persistent chemical, was often used to kill termites around foundations in the past. Pesticide mixing areas are usually 'hot spots' of contamination (Shayler et.al., 2009).

When soil is polluted with chemicals, no plants, insects or microorganisms can grow on it. As a result, the soil becomes unproductive and unfit for human habitation. Soil is necessary to keep the natural environment balanced a fertile and productive. Considering that soil is the reason we are able to sustain ourselves, the contamination of it has major consequences on human health. Crops and vegetables grown on polluted soil absorb much of the chemicals and transport them into the human body (Ganguly, 2017).

Long term exposure to polluted soil can affect the genetic information of the organism, causing congenital illnesses and chronic health problems that cannot be cured easily. In fact, it can nauseate the livestock to a considerable extent and lead to food poisoning over a long period of time. The soil contamination can even cause widespread famines if the plants are unable to grow in it (Ganguly, 2017).

The ecological balance of any system can be affected because of the widespread soil pollution. Most plants are unable to adapt when the chemistry of the soil changes rapidly and radically. The amount of fungi and bacteria found in the polluted soil that bind it together begin to decline, which creates an additional problem of soil erosion. The soil fertility slowly reduces, making land unsuitable for agricultural and any other local vegetation to survive. Soil contamination leads to large tracts of land to become dangerous to health. Unlike deserts, which are suitable for its native vegetation, such soils cannot support most forms of plant life (Ganguly, 2017).

RADIOACTIVE POLLUTION

According to Environmental Encyclopedia, radioactive pollution can be defined as the release of radioactive substances or high-energy particles into the air, water, or earth as a result of human activity, either by accident or by design. The sources of such waste include: 1) nuclear weapon testing or detonation; 2) the nuclear fuel cycle, including the mining, separation, and production of nuclear materials for use in nuclear power plants or nuclear bombs; 3) accidental release of radioactive material from nuclear power plants. Sometimes natural sources of radioactivity, such as radon gas emitted from beneath the ground, are considered pollutants when they become a threat to human health (Environmental Encyclopedia, 2005). Radioactive pollution is created when byproducts of a nuclear reaction, either man-made or natural, are dumped in the natural environment or into the artificial ecosystems.

Since even a small amount of radiation exposure can have serious (and cumulative) biological consequences, and since many radioactive wastes remain toxic for centuries, radioactive pollution is a serious environmental concern even though natural sources of radioactivity far exceed artificial ones at present (Environmental Encyclopedia, 2005).

Environmental Encyclopedia (2005) informs that nuclear power and research stations are the major contributors to man-made radioactive waste. These facilities use nuclear reaction (usually fission) for the purpose of either production of energy (electricity) or research. When a heavy atom of a nuclear fuel, such as uranium, experiences nuclear fission, it results into two daughter nuclei, both radioactive. These byproducts can't be recycled in the modern industrial processes and thus must be dumped. The environmental introduction of these radioactive byproducts causes radioactive pollution (Environmental Encyclopedia, 2005).

As Environmental Encyclopedia (2005) defines, "radioactivity is the spontaneous loss of energy from an unstable atom, in the form of various nuclear byproducts (radiation)". It helps the atom gain a relatively stable form. This spontaneous loss, known as radioactive decay, continues till a stable (nonradioactive) configuration is achieved (Environmental Encyclopedia, 2005). The reason for radiation is considered a threat is that it holds enough energy to ionize a stable atom within the organism by separating an electron from it. When ionizing radiation enters the plant, animal, or human body, it ionizes the molecules found in the organism. This leads to the formation of numerous free radicals that react with vital bodily components, nullifying their effects by forming new compounds in their place (Environmental Encyclopedia, 2005). Radioactive exposure can lead to development of cancer.

The three main types of emissions from radioactive substances are: α (alpha)-radiation, β (beta)- radiation, and γ (gamma)- radiation (Figure 51). Alpha radiation consists of helium-4 nucleus and is readily stopped by a sheet of paper. Beta radiation, consisting of electrons, is halted by an aluminum plate. Gamma radiation is eventually absorbed as it penetrates a dense material (IAEA, 2007). Lead is good at absorbing gamma radiation, due to its density. The effect of alpha particles, which are, actually, a helium atom, is the lowest and the gamma rays – is the most among the three types.

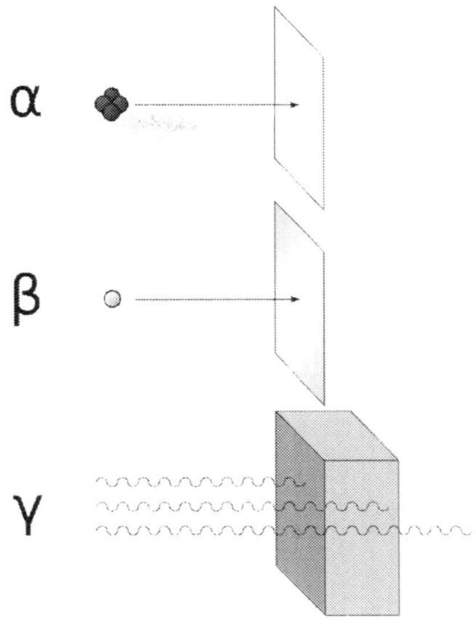

Figure 51: Three Types of Radioactive Particles.

Source: Stannered – Traced from this PNG image., CC BY 2.5, https://commons.wikimedia.org/w/index.php?curid=1716989

Alpha particles can be blocked by a simple sheet of paper or by a sufficient distance). In the same time, protection from gamma particles requires thick lead plates. However, accidental ingestion or injection of alpha particles can be fatal, since they can come in direct interaction with internal organs and important bodily fluids such as blood and ionize them (Environmental Encyclopedia, 2005).

Radioactive pollution has adverse effects on the environment and human health. International Atomic Energy Agency (2007) informs that when soil is contaminated with radioactive substances, the harmful substances are transferred into the plants growing on it. It leads to genetic mutation and impacts the plant's normal functioning. Some plants may die after such pollution, while others may develop mutant seeds. Eating any part of the plant, which underwent to radioactive exposure, primarily fruits, poses serious health risks for animals and humans (IAEA,2017).

Since plants are the base of all food chains, their exposure to radioactive substances can lead to radioactive deposition all along the food web. Similarly, when radioactive waste is dumped in a water source, it can affect the entire aquatic food web. Both terrestrial and aquatic radioactive pollution usually culminates in human consumption. Since humans are the apex predators, the accumulation of radioactive substances on the last rung of the food chain would be maximum (IAEA, 2007).

The effects of radioactive pollution on human beings can differ from mild to fatal; the magnitude of the adverse impact largely depends on the level and duration of exposure to radioactivity. Low levels of localized exposure for healthy organism usually have a superficial effect and cause mild skin irritation. Consequences of long, but low-intensity exposures include nausea, vomiting, diarrhea, loss of hair, bruises due to subcutaneous bleeding etc. (IAEA, 2007).

According to the International Atomic Energy Agency (2007), long-term exposure or exposure to high amounts of radiation can have far more serious health effects. Radioactive rays can cause irreversible damage to DNA molecules and can lead to mutations through the generation and to the life-threatening condition. Prolonged exposure ionizes a large number of molecules in the body into free radicals. Such free radicals promote the growth of cancerous cells, i.e. tumors, in the body. People with heavy radiation exposure are at a very high risk for cancers (IAEA, 2007).

The rapidly growing and dividing cells, especially skin cells, bone marrow, intestines, and gonads are more sensitive towards radioactive emissions. On the other hand, cells that do not undergo rapid cell division, such as bone cells and nervous cells, aren't injured so easily. The most common types of cancers caused by radiation are: skin cancer, lung cancer and thyroid cancer (AIEA, 2007).

The effects of genetic mutations are transmitted on to the future generations as well. For example, if the parents are exposed to nuclear radiation, their child could have severe congenital birth defects, such as physical or mental. This is tragically reported in the case of Hiroshima and Nagasaki, where the consequences of nuclear radiation were carried on for generations, and thousands of children were born with physical abnormalities and mental retardation. The radiation also brought about a spike in cancer: after more than 65 years, the region still has a much higher rate of cancer and congenital abnormalities than the rest of Japan (Hasegawa et al. 2015).

Professor Arifumi Hasegawa (2015) admits that there is no safe limit for radioactivity, since even the smallest amount has some effect on the body and holds the potential to be highly dangerous. However, radioactivity can be sensed by various devices, so that preventive measures should be taken. A Geiger counter is a commonly used device to detect radioactivity (Hasegawa et al. 2015).

International Atomic Energy Agency (2007) reasons that the effects of radioactivity can be decreased based on three ways:

- Time: Radioactive contamination of the area decreases during the time (based on half-decay period), but usually this time is much longer than human life.
- Distance: Radioactive substances owe their destructive ability to their initial velocity, which let them penetrate our skin. Travelling over long distances, particles lose velocity, and thus can't penetrate the skin and enter human body.
- Shielding: Lead is commonly used as a radiation shield in occupations where contact with radioactive substances cannot be avoided, such as nuclear power plants.

VIBRATION

"Vibration is a mechanical phenomenon whereby oscillations occur about an equilibrium point. Vibration is an oscillatory motion through a solid medium in which the motion's amplitude can be described in terms of displacement, velocity, or acceleration" (U.S. Department of Transportation, 1995). In many industrial processes vibration is undesirable, wasting energy and creating unwanted sound. For example, the vibrational motions of engines, electric motors, or any mechanical device in operation are typically unwanted. Such vibrations could be caused by imbalances in the rotating parts, uneven friction, or the meshing of gear teeth. Careful designs usually minimize unwanted vibrations.

Vibration can be a serious problem as it causes buildings to shake and rumbling sounds to be heard. In contrast to noise, vibration is not a usual environmental issue. The major sources of vibration are: construction activities (blasting, pile driving, heavy earth-moving equipment) trains, and vehicles on rough roads (U.S. Department of Transportation, 1995).

Vibration can be quantified by several different methods. According to the U.S. Department of Transportation (1995), the "peak particle velocity (PPV) is the maximum instantaneous peak of the vibration signal. The PPV is most frequently used to describe vibration impacts to buildings and is measured in inches per second. The root mean square (RMS) amplitude is most frequently used to describe the effect of vibration on the human body. The RMS amplitude is defined as the average of the squared amplitude of the signal. Decibel notation (Vdb) is commonly used to measure RMS. The decibel notation acts to compress the range of numbers required to describe vibration" (U.S. Department of Transportation, 1995).

High levels of vibration can lead to physical personal injury or damage to buildings. Most people consider vibration to be an annoyance that may affect concentration or disturb sleep. In addition, high levels of vibration may damage fragile buildings or interfere with equipment that is highly sensitive to vibration (e.g., electron microscopes). To assess the impact of vibration, the Federal Railway Administration (FRA) has published guidance relative to vibration effects. According to the FRA, fragile buildings can be exposed to vibration levels of 0.5 inches per second PPV without experiencing physical damage (Environmental Impact Report, 2008).

People are not used to experience the vibration every day as it is in the case with noise. The background vibration velocity level in living areas is limited by 50 RMS or lower, well below the threshold of perception for humans, which is around 65 RMS. Most noticeable vibration inside the buildings is caused by inside sources such as operation of mechanical equipment, movement of people, or slamming of doors. Construction equipment, steel-wheeled trains, and traffic on rough roads are the common outdoor sources of perceptible vibration. If the roadway is built properly, the vibration from traffic is rarely perceptible (Environmental Impact Report, 2008).

Craig Friedrich (1997) admitted that during the manufacturing process, aside from vibrations arising from the motions of the machine, vibration can be carried into the machine system through contact with a vibrating floor. Floor vibrations can come from a number of sources including other machine tools, elevators, walking people, outside seismic vibrations by passing trucks and trains. Normally, efforts are made to reduce the transmitted vibration to the machine tool by using pneumatic isolation, similar to that used in high quality optical tables. The isolation can be passive (simply a pressurized bladder supporting the machine tool) or more active (using servo leveling valves to maintain a level work table).

There are some methods to reduce vibration effects if the vibration source frequency is known. This can be learned by using accelerometers placed directly on the floor, or by being able to identify and quantify the vibration source. Most machine tools, because of their mass, will have a low rigid body natural frequency of perhaps only a few Hertz. This frequency is often excited by nearby walking or manual movements. For higher frequencies, there will probably be some degree of effective damping by an isolation system. This effect can be seen from the typical transmissibility chart. At higher frequency, less isolation damping is wanted to limit the transmission of the excitation. The force transmitted by the damper is proportional to the velocity of the input (as a first-order approximation) and a higher frequency displacement results in a higher velocity and a larger transmitted force to the machine. The isolation system must be carefully tuned to the dominant sources of vibration (Friedrich, 1997).

NOISE POLLUTION

The traditional definition of noise is "unwanted or disturbing sound" (Suter, 1991). According to the US EPA (1981): "sound becomes unwanted when it either interferes with normal activities such as sleeping, conversation, or disrupts or diminishes one's quality of life". The fact that you can't see, taste, or smell the noise may help to explain the fact why it has not received as much attention as other types of pollution, like atmospheric or water pollution. The air around us is constantly filled with sounds, yet most of us would probably not say we are surrounded by noise (US EPA, 2017). Though for some, the persistent and escalating sources of sound can often be considered an annoyance. This 'annoyance' can have major consequences, primarily to one's overall health.

US Environmental Protection Agency (1981) admits that noise pollution takes place when there is either excessive amount of noise or an unpleasant sound that causes temporary disruption in the natural balance. This definition is typically applicable to sounds or noises that are abnormal in either their volume or their source. It has become difficult to escape noise in our environment. Even electrical appliances at home have a constant hum or beeping sound. Lack of industrial and urban planning is the reason of increasing the exposure to unwanted sounds. That is why, understanding noise pollution is necessary to curb it in time (EPA, 2017).

Conserve Energy Future Organization describes such causes of noise pollution:

- Industrialization: Most of the industries use big machines which can produce large amount of noise. Apart from that, various equipment, like compressors, generators, exhaust fans, grinding mills also participate in producing big noise. Therefore, you must have seen workers in these factories and industries wearing ear plugs to minimize the effect of noise (CEF, 2017).
- Transportation: Large number of motor vehicles on roads, airplanes flying over cities, underground trains produce heavy noise and people get it difficult to get used to that. The high noise levels cause situations wherein a normal person loses the ability to hear properly (CEF, 2017).
- Construction Activities: Under construction activities like mining, construction of bridges, dams, buildings, stations, roads, flyovers take place in almost every part of the world. These construction activities take place every day as we need more buildings, bridges to accommodate more people and to reduce traffic congestion. The down point is that construction equipment is too noisy (CEF, 2017).
- Poor Urban Planning: In most of the developing countries, poor city design play a vital role in vibration pollution. Small apartments, large families sharing small space, fight over parking, frequent fights over basic amenities leads to noise pollution which may disrupt the environment of society (CEF, 2017).

- Social Events: Most of the social events usually produces noise at its peak. During marriage, at the parties, pub, disc or place of worship, people normally flout rules set by the local administration and create nuisance in the area. People play songs on high volumes and dance till midnight which makes the condition of people living nearby poorer. In markets, people selling food and goods via making loud noise to attract the attention of people (CEF, 2017).
- Household Chores: Humans are surrounded by gadgets and use them extensively in their daily life. Devices like TV, mobile phones, computers, mixer grinder, pressure cooker, vacuum cleaners, washing machines and dryers, coolers, air conditioners are minor contributors to the amount of noise that is produced but it affects the quality of life of your neighborhood in a bad way (CEF, 2017).

While this form of pollution may seem harmless, it in fact has harmful consequences. The adverse impact of noise pollution on the environment and human health is highly severe. Studies (US EPA, 1981; Suter, 1991; US EPA, 2017) have shown that there are direct links between noise and health. Noise pollution adversely impacts the health of millions of people. Health issues caused by noise include stress related illnesses, high blood pressure, speech interference, hearing loss, sleep disruption, and lost productivity. Noise Induced Hearing Loss is the most common and often discussed noise disease, but research has shown that exposure to constant or high levels of noise can cause countless adverse health effects (US EPA, 2017).

Effects of noise pollution can be classified in following groups:

- Hearing problems. Any unwanted sound can cause problems within the organism. Human ears can take in a certain level of sounds without getting damaged. Artificial noises such as jackhammers, horns, machinery, airplanes and even vehicles can be too high for human's hearing range. Constant exposure to high levels of noise can easily result in the damage of people ear drums and loss of hearing. It also reduces organism sensitivity to sounds that human ears pick up unconsciously to regulate our body's rhythm (CEF, 2017).

- Health issues. Excessive noise pollution within working areas such as offices, construction sites, bars and even at our homes can influence psychological health. Occurrence of aggressive behavior, disturbance of sleep, constant stress, fatigue, and hypertension can be linked to excessive noise levels. These, in turn, can cause more harmful and chronic health problems later in life (CEF, 2017).
- Sleeping disorders. Loud noise can certainly hamper human sleeping patterns and can be resulted in irritation and uncomfortable situations. Absence of a good night sleep, can cause problems related to fatigue and human performance may go down in office as well as at home. In such situations doctors recommend to have a sound sleep to give your body proper rest (CEF, 2017).
- Cardiovascular issues. Blood pressure intensity, cardio-vascular disease and stress related heart issues are on the rise because of noise. Studies suggest that high intensity noise causes high blood pressure and increases heart beat rate as it disrupts the regular blood flow. Turning noise to a manageable level depends on our understanding noise pollution and how we tackle it (CEF, 2017).
- Trouble communicating. High decibel noise can put problems in people's communication. This may lead to misunderstanding and persons may get difficult understanding each other. Constant sharp noise can lead to severe headache and disturb person's emotional balance (CEF, 2017).
- Effect on wildlife. Wildlife faces far more problems with noise pollution than human kind because they are more dependent on sound. Animals develop a better sense of hearing than people since their survival depends on it. They become disoriented more easily and face many behavioral problems. In nature, that can cause animals' suffering from hearing loss, that makes them easy prey and leads to dwindling populations. Some

animals become inefficient at hunting, disturbing the balance of the eco-system (CEF, 2017).

EPA (2017) argued that individuals can take some steps to protect themselves from the harmful effects of noise pollution. If people must be around loud sounds, they can protect their ears with hearing protection (e.g., ear plugs or ear muffs).

THERMAL POLLUTION

Thermal pollution is a temperature change in natural water bodies caused by human activity, such as use of water as coolant agent at the power plant. According to J. Henry et.al. (1996), definition of thermal pollution is "the degradation of water quality by any process that changes ambient water temperature". Thermal pollution is usually associated with higher rates of water temperatures in a stream, lake, or ocean due to the discharge of heated water from industrial processes (Pollution Issue, 2017), for example, heating process at the power plant. This activity also results in higher evaporation rates and thus, in microclimate change at the local area. The burning issue of one Nuclear power plant in Southern Ukraine which cooling process is going through the water reservoir, is that water temperature in the cooling pond never drops lower than 23 °C, that cause severe fogs and even cars accidents during the winter periods.

Increases in ambient water temperature also occur in streams where shading vegetation along the banks was removed or where physical pollution has made the water more turbid. Such actions allow more energy from the sun to be absorbed by the water and thereby increase its temperature. In some situations, the effects of colder-than-normal water temperatures may be observed (i.e., the discharge of cold bottom water from deep-water reservoirs behind large dams has changed the downstream biological communities in systems such as the Colorado River) (Pollution Issue, 2017).

The production of energy by the conversion of heat energy into mechanical energy using a heat engine (with burning fossil fuels or nuclear fission process), as well as the burning of wood in a fireplace to create heat are sources of thermal pollution. Heat engines group includes steam engines, turbines, and internal combustion engines. they work on the principal of heating and pressuring a fluid, the performance of mechanical work, and the rejection of unemployed or waste heat to a sink. According to the physical laws and their design, heat engines can only convert 30 to 40 percent of the available input energy in the fuel source into mechanical energy, and the highest efficiencies are obtained when the input temperature is as high as possible and the sink temperature is as low as possible (Hinrichs et.al., 2001).

The waste heat from power generating plants is transferred to cooling water obtained from local water bodies such as a river, lake, or ocean. Large quantities of water are used to keep the sink temperature as low as possible to maintain the high thermal efficiency (Hinrichs et.al., 2001). Roger Hinrichs and his colleagues (2001) describe the San Onofre Nuclear Generating Station between Los Angeles and San Diego, California, that has two main reactors with a total operating capacity of 2,200 MW (megawatts). Two reactors circulate a total of 2,400 million gallons per day of ocean water at a flow rate of 830,000 gallons per minute for each unit. The cooling water is brought to the station from two intake structures located 3,000 feet offshore in water 32 feet deep. The water is heated to approximately 19°F above ambient as it flows through the condensers and is dumped back into the ocean through a series of diffuser -type discharges that have a series of sixty-three exit pipes spread over 2,450 feet. This dumped water is rapidly mixed with ambient seawater by the diffusers and the average rise in temperature after mixing is less than 2°F (Hinrichs et.al., 2001).

According to Pollution Issue website (2017), the primary effects of thermal pollution are direct thermal shock, changes in dissolved oxygen, and the redistribution of organisms in the local community., Most aquatic organisms have developed enzyme systems that operate in only narrow ranges of temperature because in the natural conditions water can absorb thermal energy with only small changes in temperature. These stenothermic organisms can be killed by sudden temperature changes that are beyond the tolerance limits of their metabolic systems as it is shown at the Figure 52 (Pollution Issue, 2017).

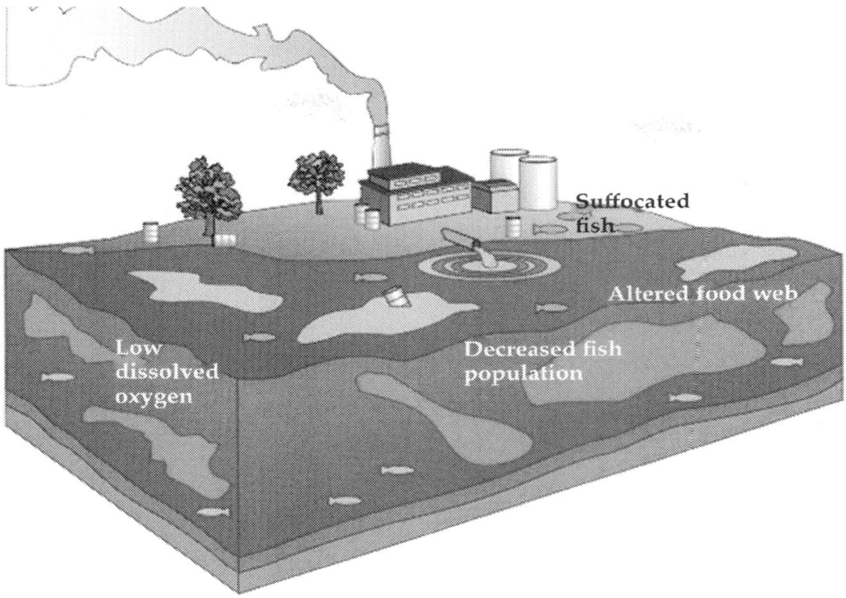

Figure 52: Ecosystem Impact of Thermal Pollution.

Source: Brooks/Colo – Thomson Learning, (2002)

The cooling water discharging by the power plants are designed to minimize heat effects on local fish communities. Though, periodic heating actions used to keep the cooling system clear of fouling organisms that clog the intake pipes can cause fish mortality (Pollution Issue, 2017). According to Brown et.al. (1983), heat treatment contraries the flow and increases the temperature of the discharge to kill the mussels and other fouling organisms in the intake pipes. Southern California Edison organization had developed a 'fish-chase' procedure in which the water temperature of the heat treatment is rized gradually, instead of rapidly, to drive fish away from the intake pipes before the temperature reaches lethal levels. The fish chase procedure has significantly decreased the rates of fish kills related to heat treatments (Pollution Issue, 2017).

Small chronic changes in temperature also adversely affects the reproductive systems of water organisms and make them more vulnerable to disease. Cold water contains more oxygen; thus, temperature rise also decrease the oxygen-carrying capacity of water. In addition, increasing the water temperature leads to rising of decomposition rate of organic matter in water, which also depletes dissolved oxygen. When these decreases in the water oxygen content occur, the metabolic rates of the aquatic organisms, which are dependent on a sufficient oxygen supply, rises because of the increasing temperature (Pollution Issue, 2017).

The configuration and diversity of ecological communities near the cooling water discharges from power plants can be adversely impacted by the direct mortality of organisms or organisms' migration from unfavorable temperature or oxygen environments. A nuclear power-generating station on Nanwan Bay in Taiwan caused decolorizing and extinction of corals in the vicinity of the discharge channel when the plant first began operation in 1988. Studies of the coral Acropora grandis in 1988 proved that the coral was bleached within two days of exposure to temperatures of 91.4°F. Coral samples taken from the same area in 1990, did not start bleaching until six days after exposure to the same temperature that shows the formation of the thermotolerance of these corals. This mechanism was enhanced by the production of heat-shock proteins that help to protect many organisms from potentially damaging changes in temperature. Some species populations can be improved by the presence of cooling water discharges as it happened with sea turtles population of in California, which is found in the southern portion of San Diego Bay near the discharge of an electrical generating station (Pollution Issue, 2017).

VISUAL POLLUTION

There are several different definitions what visual or landscape pollution is. "Visual pollution is the destruction of natural landscape view with artificial or industrial elements, which can refer to the presence of overhead power lines, motorway billboards, scarred landforms (as from strip mining), open storage of trash, municipal solid waste or space debris" (Environmental Pollution, 2010). Yilmaz (2011) defines visual pollution as "an aesthetic issue and refers to the impacts of pollution that impair one's ability to enjoy a vista or view". The term is used broadly to cover visibility, limits on the ability to view distant objects, as well as the more subjective issue of visual clutter, structures that intrude upon otherwise 'beautiful' scenes, as well as graffiti and other visual defacement (Yilmaz, 2011). Visual pollution disturbs the visual areas of people by creating harmful changes in the natural environment. Billboards, open storage of trash, antennas, electric wires, buildings, and automobiles are often considered visual pollution. An overcrowding of an area causes visual pollution. Visual pollution is defined as the whole of irregular formations, which are mostly found in natural and built environments.

Demet Yilmaz (2011) pointed out such sources of visual pollution:

- billboards;
- open storage of trash;
- space debris;
- cell towers;
- electric wires;
- buildings and automobiles.

Pollution Issue website (2017) defines that visibility is a measure of how far and how well people can see into the distance. Haze obscures visibility. It is caused when light is absorbed or scattered by pollution particles such as sulfates, nitrates, organic carbon compounds, soot, and soil dust. Nitrogen dioxide and other pollution gases also contribute to haze. Haze increases with summer humidity because sulfate and other particles absorb moisture and increase in size. The larger the particles, the more light they dissipate (Pollution Issue, 2017).

Haze is not only the urban type of visual pollution, it has dramatically seen as a brownish-grey cloud obscuring many beautiful vistas in natural areas. National Research Council Board on Environmental Studies and Toxicology (1991) informed, that at Acadia National Park in Maine, visual range on a clear day can be 199 miles. On a hazy day, visual distance be reduced to 30 miles. In some conditions, haze at Grand Canyon National Park was so severe that people could not see across the 10-mile wide canyon. The source of this landscape pollution is huge coal-fired electric plant, the Navajo Power Generating Station, situated 80 miles north of the Grand Canyon. To prove this in 1985, researchers from Colorado State University injected methane-containing deuterium into the power plant's smoke emissions. Deuterium is not normally present in the air. When monitors determined the presence of deuterium in canyon air, researchers could demonstrate that the plant was responsible for much of the canyon haze. To solve this problem, Navajo's owners agreed to a 90-percent cutback in sulfur dioxide emissions by 1999. In addition, the U.S. Environmental Protection Agency (EPA) has issued regional haze regulations that call on states to create goals and strategies and on working together in regional groups to improve visibility in 156 national parks and wilderness areas (NRCBET, 1991).

Visual blight—billboards, power lines, cell towers, even ugly buildings—is literally in the eye of the beholder. It is subjective. To the industrial manager, a well-placed billboard may be a thing of beauty. But to the traveler whose view of the rolling hills or the rustic village is blocked, it is visual pollution (Pollution Issue, 2017).

As Gudis (2003) describes in his article, billboards proliferated in the 1940s and 1950s, spurred by the growth of automobile traffic and construction of interstate highway system, but in 1965 Lady Bird Johnson, wife of President Lyndon Johnson, attacked their growing occurrence on our nation's roadways. "Ugliness is so grim," the first lady declared, and she fought for and won passage of the Highway Beautification Act of 1965. This groundbreaking law prompted a number of states, including Alaska, Hawaii, Maine and Vermont, to ban billboards totally (Gudis, 2003).

Effects of exposure to visual pollution include (Yilmaz, 2011):

- Distraction;
- Eye fatigue;
- Decreases in opinion diversity;
- Loss of identity.

In some communities, sensitivity to visual pollution has led utility companies to put power and telephone lines into the ground. A decade ago wind farms were considered as visual polluters in Eorupe, but today we can't imagine German landscape without windmills. The latest fight against visual pollution concentrates on cell towers, needed to provide cellular telephone service. One solution has been to disguise cell towers as trees or cacti, or farm grain storage. Graffiti, spray-painted names, and messages, are a form of urban visual blight. Attempts to curb graffiti by banning the sale of spray paint to minors have had little effect (Pollution Issue, 2017).

EFFECTS ON BIOTA

As we have analyzed previously, industrial pollution causes natural rhythms and patterns to fail, meaning that the wildlife is getting affected in a severe manner. Habitats are being lost, species are becoming extinct and it is harder for the environment to recover from each natural disaster. Major industrial accidents like oil spills, fires, leak of radioactive material and damage to property are harder to clean-up as they have a higher impact in a shorter span of time (CEF, 2017).

Pollution may ruin landscapes, poison soils and waterbodies, and even kill plants and animals. Heavy metals and other toxins, migrating through the food chains, accumulate in top predators and make some species unsafe to eat. World Wildlife Fund (2017) explains that when toxic chemicals and metals enter the environment, organisms may absorb them through their skin or ingest them in their food or water. Animals higher in the food pyramid accumulate these toxins in higher and higher concentrations (every next level has 10 times higher concentration than previous). This process is called biomagnification. Top predators—including fish, birds, and mammals—can have much higher levels of these toxins in their bodies, making them more likely to experience the diseases, birth defects, genetic mutations, and other deleterious effects of these poisons (WWF, 2017).

Implementation of pesticides and fertilizers on farm lands has increased by 26-fold over the past 50 years, increasing the crop production globally. Killing weeds and target insects, pesticides can also kill non-target organisms, including beneficial insects, soil bacteria and fish. Fertilizers are not directly toxic, but their presence can destroy the nutrient system in freshwater and marine areas. This change can result in eutrophication – an explosive growth of algae due to excess nutrients. As a result, the water is depleted of dissolved oxygen, and fish and other aquatic life may be killed (WWF, 2017).

Light and noise pollution also impact the wildlife. According to the WWF (2017), in the Arctic, the oil and gas exploration equipment sounds so loud that belugas, bowhead whales and other sea life have had difficulty feeding and breeding. Artificial light disrupts circadian rhythms for both humans and animals alike and may even contribute to the development of cancer. Light pollution also can impact sea turtles: adult and hatchling sea turtles swim toward lights along the beach, thinking they are heading toward the moon. Coastal developments, therefore, are stimulated to turn off their lights or cover them at night (WWF, 2017).

Noise pollution affects wildlife more severe than humans because they are more dependent on sound. Pollution Issue website (2017) explains that animals develop a better sense of hearing than us since their survival depends on it. Animals become disoriented more easily because of noise impact and face many behavioral problems. In nature, animals may become inefficient at hunting, disturbing the balance of the ecosystem (WWF, 2017).

Wild species that depend on mating calls to reproduce are often unable to hear these calls because of excessive artificial noise. As a result, they are unable to mate that causes the declining in populations. Other animals produce special sound waves to echo-locate and find their way when migrating. When human activity disturbs their sound signals these animals get lost easily and do not migrate when they should. To cope up with the cumulative sound around them, animals are even becoming louder, which may further add to the pollution levels (Pollution Issue, 2017).

Bignal et al. experimented on the effect of vehicle emissions and bryophyte plants near roadways. Near highly used roads there were more nitrogen-tolerant plants than in areas further from the road; vehicle emissions release nitrogen into the atmosphere. The ions emitted by vehicles do not last long in plumes in the air. Near busy roads some of the exhaust falls to the ground in soot particles, which can affect the plants, (Jayaratne et al., 2010). As noted in the Bignal et al. experiment the bryophytes were not found in highly urbanized areas close to pollution sources, because of the high concentrations of smoke and sulfur dioxide, (2008). Even if the plants still exist in the locations near highly traveled roads they, similarly to humans, have increased susceptibility to environmental stresses, (Bignal et al., 2008).

- R. Shah, (2016) describes such effects of air pollutants on plants:
- NO_2 results in necrosis (dead areas on a leaf), chlorosis (reduction of chlorophyll), epinasty (downward curvature of

leaf) and abscission (dropping of leaves);

- SO_2 causes bleached spots, chlorosis and affects photosynthesis along with ozone and PAN;
- CO at higher levels (100 to 1000 ppm) causes leaf curling, leaf drop, reduced leaf size and interference with cellular respiration;
- PAN results in silvering of lower surface of leaf, damage to sensitive leaves and suppressed growth (Shah, 2016).

Air-borne pollutants accumulate in forage and poison the animals when they graze it. Live-stocks near smelters suffer arsenic poisoning with symptoms like salivation, thirst, and liver necrosis (Shah, 2016).

The ecological balance of any system gets affected due to the widespread contamination of the soil. Most plants are unable to adapt when the chemistry of the soil changes so radically in a short period of time. Fungi and bacteria found in the soil that bind it together begin to decline, which creates an additional problem of soil erosion. The fertility slowly diminishes, making land unsuitable for agriculture and any local vegetation to survive. Soil pollution can negatively affect the metabolism of microorganisms and arthropods, which can destroy some layers of the primary food chain and have a harmful effect on predator animal species. Also, small life forms may consume harmful chemicals in the soil which may then be passed up the food chain to larger animals, which may lead to increased mortality rates and even animal extinction (Everything Connects, 2013).

Agricultural production, as well as other human activities, have led to large increases in the levels of nitrogen and phosphorus in the environment. In water, this excess of nutrients, a process called eutrophication, can fuel the extreme growth of phytoplankton and algae, which can sometimes have shocking consequences. Harmful algal blooms – blooms of species that produce deadly toxins and sometimes known as 'red tides' or 'brown tides' for their attendance in the water – can kill fish, marine mammals and seabirds and influence the human health. After the algae and other organisms that had been allowed to bloom because of the nutrient excess eventually die off, bacteria may suck up all the oxygen from the water in the process of algae decomposition. This hypoxia creates a 'dead zone' where water animals cannot live. More than 400 areas around the world have been recognized as suffering from eutrophication and 169 – from hypoxia (WWF, 2017).

Plastics and other marine trash floating on the water surface may persist in the oceans for years, accumulating in the centers of ocean gyres, creating great 'garbage patches'. This patches are areas where concentrations of flotsam and jetsam, mostly small pieces of plastic, are extremely high. This litter can disperse toxic chemicals throughout the oceans, snag and tear corals, and harm animals if they ingest pieces of plastic or become entangled in the debris (WWF, 2017).

EFFECTS ON HUMAN HEALTH

According to the Rio-Declaration (UN, 1992) and to the theory on Environmental safety, humans are in the center of care in the process of sustainable development. Thus, we should look at all of environmental effects through the glass of possible human health effects.

Ecosystems provide their special services to people. Ecosystem services are indispensable to the wellbeing of all people, everywhere in the world. They include provisioning, regulating, and cultural services that directly affect people, and supporting services needed to maintain the other services. The World Health Organization (2017) admits that human health and well-being depends on these services, from the availability of adequate food and water, to disease regulation of vectors, pests, and pathogens.

The causal links between environmental change and human health are complex because they are often indirect, displaced in space and time, and dependent on a number of modifying forces. Human health ultimately depends upon ecosystem products and services (such as availability of fresh water, food, and fuel sources) which are requisite for good human health and productive livelihoods. Significant direct human health impacts can occur if ecosystem services are no longer adequate to meet social needs. Indirectly, changes in ecosystem services affect livelihoods, income, local migration and, on occasion, may even cause political conflict. The resultant impacts on economic and physical security, freedom, choice, and social relations have wide-ranging impacts on well-being and health, and the availability and access to health services and medicines (WHO, 2017). Figure 53 shows us the complex system of interrelations between ecosystem services, environmental impacts, and human health.

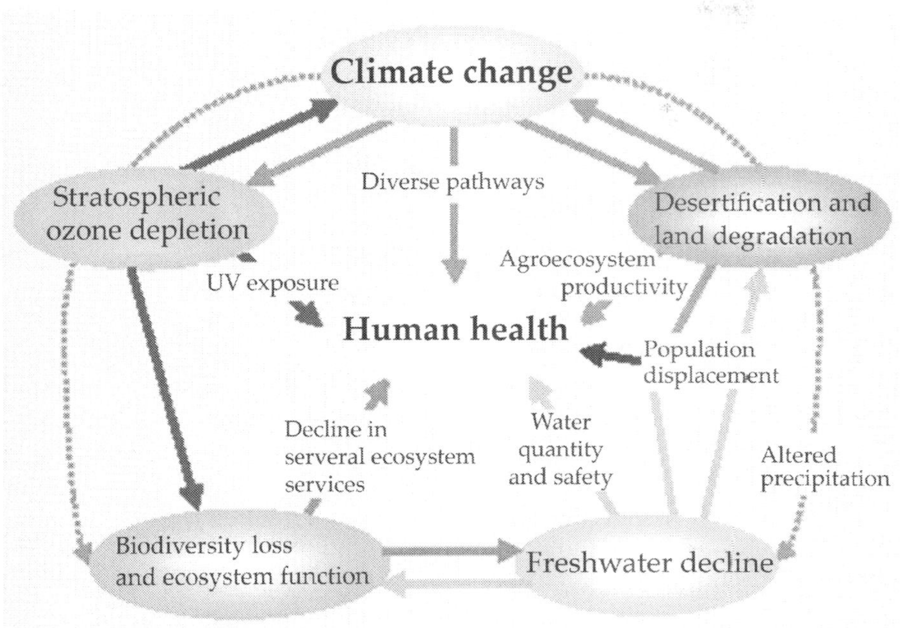

Figure 53: Interrelations Between Environmental Impacts and Human Health.

Source: WHO (2017). Available at: http://www.who.int/globalchange/ecosystems/en/

Ecosystem disruption can impact on human health in a variety of ways and through complex pathways. The types of health effects experienced are determined by the level to which population is depended on ecosystem services, and social factors such as poverty, access to food and water.

Wisdomjobs (2012) tried to classify effects of environmental pollution on human health:

I. Effects of Air Pollution

- Impact on Lung functioning
- Asthma
- Itching of eyes, mouth and throat

- Respiratory disorders
- Cough and wheezing
- Reduced energy levels
- Headache
- Negative impact on reproductive and immune systems
- Neurological disorders
- Cancer

II. Effects of Water Pollution:

- Waterborne diseases like Typhoid, Amoebiasis, Giardiasis, Ascariasis and Hookworm diseases
- Rashes, earache, and pink eye
- Hepatitis, encephalitis, gastroenteritis and vomiting, stomach aches and diarrhea
- Prostate cancer
- Lack of developmental growth
- Liver and kidney damage
- DNA damage
- Exposure to mercury causes Parkinson's disease

III. Effects of soil pollution

- Leukemia
- Effects children's brain and causes developmental damage
- Kidney damage
- Damage to central nervous system
- Headache, fatigue, skin rash and eye irritation
- Contamination of crop brings up problems like food security

IV. Effects of Noise pollution

- Noise pollution decreases the human efficiency
- Reduced concentration
- Fatigue
- Fertility problems in women
- Increased blood pressure
- Temporary/Permanent deafness (Wisdomjobs, 2012).

Figure 54 presents schematic view of health effects of pollution.

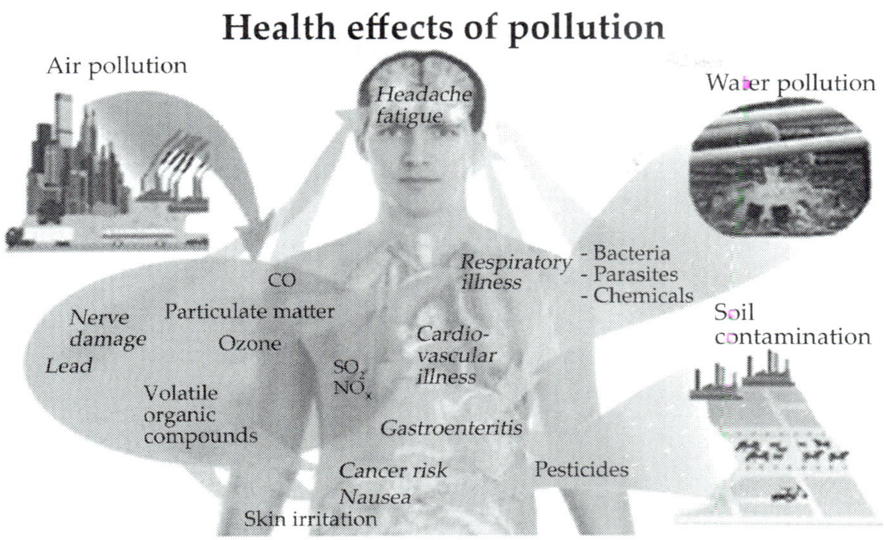

Figure 54: Heath Effects of Pollution.

Source: Wisdomjobs. (2012). Available at: https://blog.wisdomjobs.com/pollution-its-causes-impact-and-prevention/

According to Shah (2016), acute exposure of NO_2 in man causes bronchiolitis fibrosis obliterans, emphysema, lung cancer, pneumonia and silofilter's disease. $NO_2$is also known to be transformed into methemoglobin in blood. SO_2, a severe allergenic agent, may cause watery nasal discharge, sneezing, dyspnea, desquamation, and bronchial spasms. CO, when inhaled, passes through the lungs and diffuses directly into blood stream where it combines with hemoglobin to form carboxy hemoglobin. CO has affinity for hemoglobin, 210 times more than O_2. Hemoglobin is thus unable to transport O_2 to body tissues causing asphyxiation (lack of O_2). Its long exposure may cause dizziness, unconsciousness and even death. Particulates with SO_2 cause silicosis, bronchitis and increased absenteeism among older workers (Shah, 2016).

Considering how important soil is to sustain people life, the contamination of it has major consequences on our health. Crops and plants grown on polluted soil absorb much of the pollution and then pass these on to us. This could explain the sudden surge in small and terminal illnesses. Long term exposure to such soil can affect the genetic make-up of the body, causing congenital illnesses and chronic health problems that cannot be cured easily. In fact, it can sicken the livestock to a considerable extent and cause food poisoning over a long period of time. The soil pollution can even lead to widespread famines if the plants are unable to grow in it (Everything Connects, 2013).

5 CHAPTER

ENVIRONMENTAL IMPACT OF INDUSTRIAL SECTORS AND METHODS OF ITS ELIMINATION

MINING INDUSTRY

Mining is an industrial sector that extracts valuable minerals or other geological materials from the earth crust6 usually from an orebody, lode, vein, seam, reef, or placer deposits. Ores recovered by mining include metals, coal, oil shale, gemstones, limestone, chalk, dimension stone, rock salt, potash, gravel, and clay. Mining is required to obtain raw materials that cannot be grown through agricultural processes, or created artificially in a laboratory or factory. Mining includes extraction of any non-renewable resource such as fossil fuels, or even water.

Mining technologies can be divided into two common excavation types: surface (open) mining and sub-surface (underground) mining. Mining operations usually create a negative environmental impact, both during the mining activity and after the mine has closed (Figure 55).

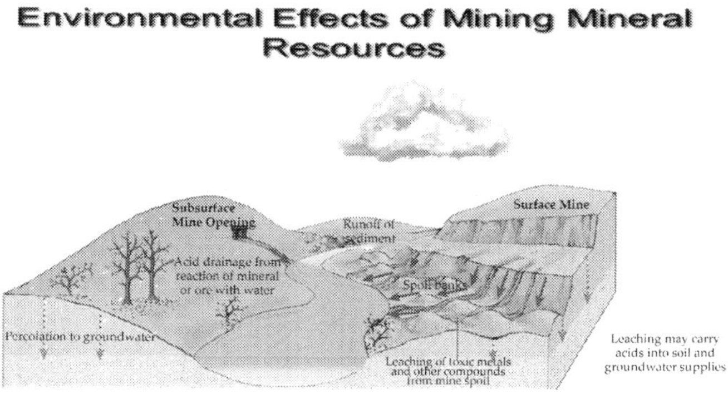

Figure 55: Environmental Effects of Mining.

Source: Brooks/Cole – Thomson, (2005).

One of the most common forms of mining is open pit mining, when raw material is excavated from an open pit. It is usually used for strategic minerals. This type of mining is extremely damaging to the environment because strategic minerals are often only available in small concentrations, which rises the amount of ore needed to be mined. Environmental impacts are present during every step of the open-pit mining process. Hard rock mining takes rock from the ground that has lain unexposed for geological eras. When smashed, these rocks expose radioactive elements, asbestos-like minerals, and metallic dust into the atmosphere. During separation, residual rock slurries, which are mixtures of pulverized rock and liquid, are produced; and tailings, toxic and radioactive elements from these liquids can enter the bedrock or ground water if not properly contained (Betournay, 2011).

Another type of mining activity is underground mining that has the potential for tunnel collapses and land subsidence (Betournay, 2011). It causes large-scale movements of waste rock and vegetation, like open pit mining. Additionally, underground mining can release toxic substances into the air and water bodies. As water used during underground mining process takes on harmful concentrations of minerals and heavy metals, it becomes a pollutant. This contaminated water can pollute the area surrounding the mine and beyond as it is migrating through. Most underground mining operations increase sedimentation in nearby rivers through their use of hydraulic pumps and suction dredges; attacking with hydraulic pumps removes ecosystem valuable topsoil containing seed banks, making it difficult for vegetation to grow again (Miranda et al., 1998). Deforestation caused by mining processes can also cause the disintegration of biomes and contributes to the effects of erosion.

In situ leach (ISL) mining process, when the ore body is dissolved and then pumped out, leaving minimal surface disturbance and no tailings or waste rock, has some environmental and occupational safety advantages over conventional mining (World Nuclear Association, 2012). There is no ore dust or direct ore exposure to the air and a lower consumption of water is needed in the ISL mining process. On the other hand, the strong acid solutions are used to dissolve the ore body and this commonly dissolves metals in the host rock as well. The fluids that remain after the leaching process usually contain high concentrations of metals and radioactive elements, posing a significant risk to nearby ground and surface water sources. Additionally, the high acidity of in situ leach mining wastewater can result in acidification of the surrounding environment (IAEA, 2005).

Environmental impacts of heap leaching are concentrated on the failure to keep process solutions within the heap leaching circuit. Toxic heap leaching fluids release into the environment can influence the surrounding ecosystem and human health (Reichardt, 2008). Water balance is crucial in heap leaching projects due to the possibility of the overflow of solutions containing toxic concentrations of heavy metals after a heavy rainfall or rapid snowmelt (Norman and Raforth, 1994). Eisler (1991) reported that in some cases, cyanide is used to deoxidize metals from oxidized ores and the resulting leach ponds can cause significant wildlife mortality. The fact of about 7,613 animal deaths between 1980 and 1989 was recorded at cyanide-extraction ponds in California, Nevada, and Arizona (Eisler, 1991).

Brine mining involves the process of "extracting and evaporating the brine solutions to remove harmful elements and compounds, potentially releasing them into the environment" (Gruber et al., 2011). During the process of drilling and transportation of brine solutions existing ecosystems can be damaged. High salinity content of brine solutions can cause unexpected leaks from well casings, pipelines, and storage tanks, which are subject to corrosion. Such leaks usually result in contamination of adjacent bodies of water (New York State Division of Mineral Resources, 1988).

Environmental impacts on air quality: All mining technologies impact the quality of air, as unrefined materials are released to the surface when mineral deposits are exposed from the site. Mining requires soil to be removed from the mining surface. Along with soil vegetation is also removed that eventually exposes the soil to weather, causing the particles to become airborne through road traffic and wind erosion. The crude particles can be composed of toxic materials such as lead, cadmium, arsenic, other heavy metals, or radioactive substances. Such particles can cause human health effects contributing to illnesses related to the respiratory systems such as emphysema. In addition, such particles can also be absorbed through the skin or ingested causing skin and stomach diseases (Worldatlas. 2016).

Land effects: Worldatlas (2016) reports that mining can cause physical destruction to the surrounding land by creating landscape blots such as open pits and piles of waste rock (Figure 56).

Figure 56: Piles of Waste Rock in Donetstk Coal-Mining Region, Ukraine.

Source: Donbass Terrikons. Available at: http://terrikon.donbass.name/uploads/posts/2012-06/1340477249_t.-chumakovskaya.jpg

Such 'terrikons' are very flammable, especially during hot summer time, polluting the air with products of combustion. In addition, they cause visual pollution.

After the miming is finished, a huge possibility exists that many of the previous landscape features cannot be renovate. Landscape declination as a result of ground movements on the earth surface can cause damage to roads and buildings because of collapsing overlaying sheets. Worldatlas

(2016) tells us about 500 subsidence collapse features contributed to the abandonment of underground metal mines which were discovered in the Galena, Kansas area between the years of 1980 to 1985 where lead ore mining took place from 1850 to 1970. The whole mining area was later re-cultivated between 1994 and 1995 (Worldatlas, 2016).

Water pollution: Water pollution caused by mining can include metal contamination, increased sediment levels in streams, and acid mine drainage. Major sources of water pollution in mining industry include processing plants, tailing ponds, underground mines, waste-disposal areas, active or abandoned surface or haulage roads. The sediments are most likely from higher levels of soil erosion causing siltation or the smothering of stream beds. Mining results in abnormally high concentrations of chemicals such as mercury, arsenic and sulfuric acid over a large area surrounding the mine. Salinization caused by mining affects irrigation, swimming, fisheries, domestic water supply, and other uses of streams (Worldatlas. 2016).

The United States has enforced federal and state mining laws demanding companies to meet and adhere to the standards put in place for the surface and groundwater protection from contamination (Paul and Campbell, 2011). If the mining site is already contaminated, then mitigation techniques such as acid mine drainage must be performed. The five main techniques are pointed out to be used in the monitoring and controlling of water flow at mining sites. They include subsurface drainage systems, groundwater pumping systems, diversion systems, subsurface barriers, and contamination ponds. During the acid mine drainage process, polluted water is pumped into the treatment facility that finally neutralizes the contaminants (Worldatlas. 2016).

Molycorp, (2012) admits that most modern mining techniques have high water demands for extraction, processing, and waste disposal. Wastewater from these processes can contaminate water bodies and streams nearby and deplete freshwater supplies in the region surrounding the mine. Some mines, such as the Mountain Pass mine in southern California, have implemented waste-water recycling technologies, resulting in a significant decrease in freshwater demands and in amount of sewage water (Molycorp, 2012).

Effects on biodiversity: Destruction or drastic modification of the original landscape and toxic substances release can have major impact on biodiversity in the area. Devastation of the habitat is the main component of biodiversity losses, but direct poisoning caused by mine-extracted elements, and indirect poisoning through food and water, can also affect animals, vegetation, and microorganisms. Habitat modifications such as pH and temperature modifications disturb communities in the area (Maest et.al., 2006). Endemic species are especially vulnerable, since they need very specific environmental conditions. Destruction or slight modification of their habitat endanger them with the risk of extinction. Habitats can be destroyed when there is not enough terrestrial as well by non-chemicals products, such as large rocks from the mines that are discarded in the surrounding area with no concern for impacts on natural ecosystem (Nicolet Minerals Company, 1998).

Effects on biodiversity are known to decrease with distance from the mine; impacts can vary greatly depending on mobility and bioavailability of the contaminant: less-mobile molecules will stay inert in the environment while highly mobile molecules will easily move into another compartment or be taken up by organisms. Bioaccumulation plays an important role in polluted habitats: mining impacts on biodiversity should be assessed if concentration levels are greater in the species on the top of the food chain because of this phenomenon (Maest et.al., 2006).

Adverse mining effects on biodiversity depend on the nature of the contaminant, the level of pollution concentration in the environment, and the nature of the ecosystem itself. Some species can have special resistance to anthropogenic disturbances, while some others will completely disappear from the contaminated zone. Just time period is not enough to allow the habitat to recover completely from the contamination (Maest et.al., 2006), and in most of the cases remediation will not enable the recovery of the diversity present before the mining activity (Lehrke, 2004).

The mining industry can impact aquatic biodiversity in different ways. Direct biota poisoning is the first way and risks are higher when pollutants are mobile in the sediment or bioavailable in the water (Kuipers et.al., 2006). Mine drainage can change the water pH. In such cases, it is hard to differentiate direct impact on organisms from impacts caused by pH changes as effects can nonetheless be observed and proved to be caused by pH modifications (Lehrke, 2004). Mining process can also affect aquatic organisms in physical way. For example, streams with high concentrations of suspended sediment limit light, thus weakening algae biomass. Metal oxide presence can limit biomass by coating algae or their substrate, thereby preventing colonization (Maest et.al., 2006).

Soil texture and water content can be greatly modified in disturbed sites, leading plants communities to be changed in the area. Most of the plants have a low level of tolerance for metals in the soil, but sensitivity differs among species. Grass diversity and total grass cover is usually less impacted by high contaminant concentration than forbs and shrubs (Maest et.al., 2006). Mines waste-material rejects or traces due to mining activity can be found near the mine, sometimes far away from the source. Established plants are not able to move away from perturbations, and can eventually die if their habitat is contaminated by heavy metals or metalloids at concentration too high for their physiology. Some species are more resistant and will survive these levels, and some invasive species that can tolerate these concentrations in the soil, will migrate in the mine surrounding lands to occupy the ecological niche (Schuler et.al., 2011).

Cultivating crops might be a problem near mines. Most crops can grow on weakly contaminated sites, but yield is generally lower than it would have been in regular growing conditions. Plants also tend to accumulate heavy metals in their aerial organs, possibly leading to human intake through fruits and vegetables. Regular consumptions of such products might lead to health problems caused by long-term metal exposure. Cigarettes made from tobacco planted into the contaminated soils might as well have adverse effects on human health, as tobacco tends to accumulate cadmium and zinc in its leaves (Maest et.al., 2006).

Animal habitat destruction is one of the main issues of mining activity. Huge areas of natural habitat are destroyed during mine construction and exploitation, forcing animals to leave the site. Animals can be poisoned directly by mine products and residuals. Heavy metals, accumulated in the plants or the smaller organisms can also lead to animal poisoning through food chains: horses, goats and sheep are exposed in certain areas to potentially toxic concentration of copper and lead in grass (Schuler et.al., 2011). Fewer number of ants species in soil containing high copper levels, in the vicinity of a copper mine, was recorded (Maest et.al., 2006). If fewer ants are found, chances are great that other organisms leaving in the surrounding landscape are strongly affected as well by this high copper levels, since ants are a good environmental control: they live directly in the soil and are thus pretty sensitive to environmental disruptions. Microorganisms are extremely sensitive to environmental modification because of their size. Such factors as modified pH, temperature changes or chemicals concentration of contaminants are especially influencing. For example, the presence of arsenic and antimony in soils led to a diminution in total soil bacteria (Kuipers et.al., 2006). Moreover, as in water, a small change in the soil pH can provoke the remobilization of contaminants (Lehrke, 2004), in addition to the direct impact on pH-sensitive organisms (Maest et.al., 2006).

Mining processes also affect human health. There are many diseases that can be caused by pollutants that are released into the air and water during the mining process. For example, during ore extraction operations enormous quantities of air pollutants, such as the suspended particulate matter, SOx, arsenic particles, and cadmium are emitted. Metals are usually emitted into the air as particulates (Maest et.al., 2006).

There are also many occupational health risks. According to Lehrke, (2004), most of the miners suffer from various respiratory and skin diseases. Miners working in different types of mines undergo to asbestosis, silicosis, black lung disease etc. (Lehrke, 2004). Human activity and households are also affected by the occurrence of landslides, floods, etc.

Specific contaminant materials

Radionuclides. According to Paul and Campbell (2011), all rare-earth-element-bearing minerals contain low levels of the radioactive isotopes that can become concentrated in mine tailings. Radioactive substances are released as dust during mining or from exposed waste rock stockpiles where they are least containable (and mostly airborne). Radionuclides can also enter the ground, and close water sources after they have been separated into tailings, if the tailings are not stored safely. Once radiation is in an ecosystem, it accumulates in plants, where the higher concentrations are ingested and ascend the levels of the food chain (Paul and Campbell, 2011). Radioactive pollution has become such a problem that monazite mining was turned off after China and the United States had imposed strict regulations (Schuler et.al., 2011).

Dust and metal. When companies destroy or explode materials during mining, the dust can release a variety of heavy metals commonly associated with health problems. As dust, these minerals (such as the asbestos-like mineral riebeckite) can be absorbed into lung tissue, causing problems like pneumoconiosis and silicosis, commonly known as 'black lung' (Paul and Campbell, 2011). Another example of harmful dust generated is flue dust, a byproduct of mining fluorine. According to the Chinese Society of Rare Earths, every ton of rare-earth element produced generates 8.5 kg of fluorine and 13 kg of flue dust, waste materials which contain the heavy metals discussed above (Schuler et. al., 2011).

Carbon output. Mining, like most heavy industries, is dependent on oil and other fossil fuels, which generate the energy needed to operate a mine. To fight these carbon emissions, some countries have enacted regulations requiring emission credits, but many countries do not have codes dealing with carbon output (Molycorp, Inc., 2010).

As we can see, nowadays mining industry faces many environmental issues that require technical solutions that will influence the entire network of economic, social, and environmental interrelations. Mining industry needs a new image based on innovative technologies which can exploit minerals at greater depths and up-to-date economic structure to support high productivity and safe working settings.

According to the Massachusetts Institute of Technology report, new mining technologies and regulations have significantly improved mining efficiency and reduced environmental impact in recent years. In general, mining techniques become much more environmentally sensitive when efficiency is improved because less waste is produced. The plan for improving efficiency and decreasing the environmental impact of mining is broken up into the following categories (MIT, 2015):

- Shutting down illegal and unregulated mines
- Choosing environmentally friendly general mining processes
- Implementing recently discovered green mining technologies
- Cleaning up the sites of shut-down mines
- Reevaluating Cut-off Grades
- Research and Development of Green Mining Technology

The I²MINE project (2013), funded by the European Commission, marks the start of a series of development activities aimed at realizing the concept of an invisible, zero impact, deep mine. The extractive sector, still seen as being old fashioned and environmentally unfriendly, will join forces to review this image by showing that mineral extraction and processing can be proceeded in a highly innovative and sustainable manner with low impact underground and zero impact above ground (I^{2}Mine, 2013).

The strategy of I^{2}Mine project is to develop innovative the methods, technologies, machines, and equipment necessary for the efficient exploitation of minerals and disposal of waste, both of which will be carried out underground. This will intensively reduce the amount of surface transportation of minerals and waste, minimizing above ground installations, and reducing the environmental impact (I^{2}Mine, 2013).

New eco-efficient technologies will be applied to make the entire mining process more efficient and environmentally sound. The concept of Intelligent Green Mine (Figure 57) is based on integrated mining process, with most of the installations underground, and only the final product will be transported above ground to be shipped to the customer. Production waste is treated and stored underground and gaseous emissions are managed underground as far as practical (I^{2}Mine, 2013). Green Mine Project focuses on ideas and concepts that increase energy efficiency and decrease waste.

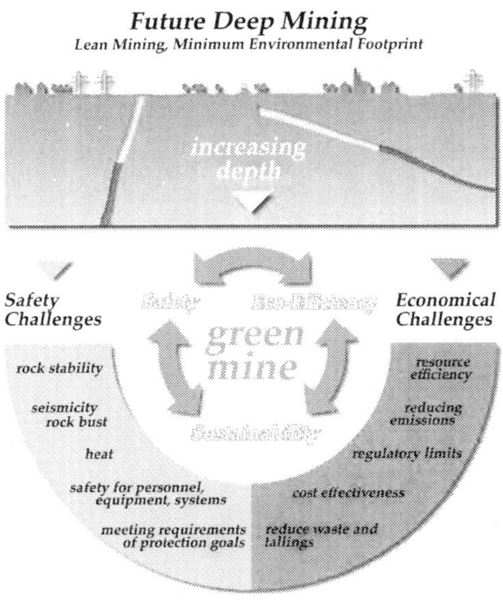

Figure 57: The Concept of Intelligent Green Mine.

Source: I²Mine. (2013). Innovative Technologies and Concepts for the Intelligent Deep Mine for the Future. Available at: http://www.i2mine.eu/content/open_access/introduction

In deep mining, health and safety aspects are very important. The ultimate safety in deep mining can only be achieved when no humans are present in the operational area (concept of zero entry). Even though not immediately achievable, the zero entry mine will be designed based on full automation, semi-automation, and remote control for all machinery. The project develops innovative solutions in terms of methods, technologies, machines, and equipment, and it is based on key concepts of Environmental Ecology, such as Life Cycle Analysis, Closed-Loop Systems, Decarbonization, Eco-Efficiency (I²Mine, 2013).

The core of the project is breakthrough technology for autonomous, highly selective, continuous mineral extraction processes and machinery. It is based on new sensor knowhows, face front separation as well as innovative concepts for mass flow management and transportation integrating state of the art technologies. The concept of an invisible, zero impact mine requires organizing the refinery process underground that selectively extracts the mineral thereby reducing waste (I²Mine, 2013). To achieve this objective, improved extraction machines and near to face processing methods, including backfill procedures, have to be developed. These innovative technologies include rock mechanics and ground control solutions, incorporating health, safety, and environmental issues. Figure 58 shows us project methodology.

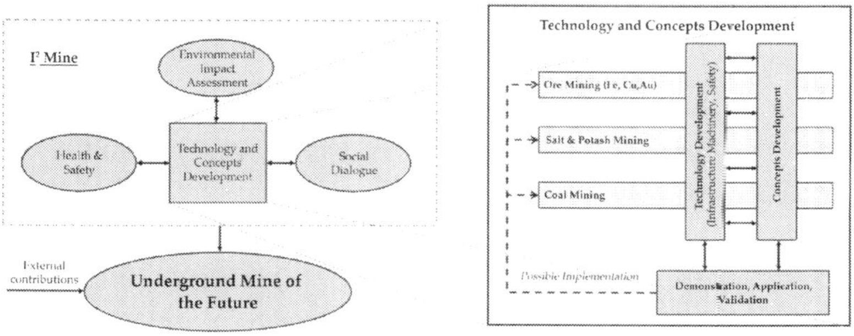

Figure 58: I²Mine Methodology Scheme.

Source: I²Mine. (2013). Innovative Technologies and Concepts for the Intelligent Deep Mine for the Future. Available at: http://www.i2mine.eu/content/open_access/introduction

I^{2}Mine focuses on the entire life cycle of mining with the following objectives (I^{2}Mine, 2013):

- Concepts for innovative mining approaches for deep deposits (steep and flat), leading to upgraded resource efficiency through higher extraction rates (20% and more), a higher selectivity of extraction (10 to 20%) together with a 20% higher deposit utilization as well as improved productivity and decreased production costs by 20%;
- Tailor-made technologies for underground, near-to-face processing to reduce the mass flow to the surface by 15 to 25%;
- High resolution 3D exploration devices for deep deposits;
- New concepts for mine running to reduce operational expenditure and new approaches for predicting, monitoring, and controlling subsidence;
- New methods to handle waste rock underground and backfill products with similar characteristics to the original rock;
- Health and safety design criteria and instructions for the new concepts and technologies developed;
- Concepts for clean, safe, and comfortable climate conditions in the underground workings;
- New concepts and technologies for deep mine rescue;
- On-line best practice database for all environmental solutions and problems associated with deep mining projects (water management, waste management, emissions, subsidence, etc.) (I2Mine, 2013).

Presented concept is expected to improve the competitiveness of the extractive sector in Europe as well as the design, manufacture and supply of equipment (I2Mine, 2013).

ENERGY SECTOR

U.S. Environmental Protection Agency defines electricity system as "a complex network of power plants, transmission and distribution wires, and end-users of electricity". Today, most US citizens receive their electricity from centralized power plants using a wide variety of energy resources to produce electrical energy, such as coal, natural gas, nuclear energy, or renewable resources such as water, wind, or solar energy. This multipart system of generation, transportation, and end-users is called electric power grid (EPA, 2017).

If we analyze the electricity system of the United States, we can see that electricity there is generated using a variety of resources. The three most common are coal, natural gas, and nuclear power. Some of the fastest growing sources are renewable resources such as wind and solar. Centralized power plants generate most of the U.S. electricity. A much smaller but growing volume of electricity is produced through distributed generation—a variety of technologies that generate power at or near where it will be used, such as onsite solar panels and combined heat and power (EPA, 2017). Figure 59 shows us the net electricity generation in the U.S. by source.

Once electricity is generated at a centralized power plant, it is sent through a series of interconnected, high-voltage transmission lines to customers. Substations 'step down' high-voltage power to a lower voltage, sending the lower voltage electricity to the end consumers through a network of distribution lines (EPA, 2017). Environmental Protection Agency (2017) informs that today in the developed countries residential, commercial, and industrial customers each account for roughly one-third of the nation's electricity use. The transportation sector accounts for a small fraction of electricity use (EPA, 2017).

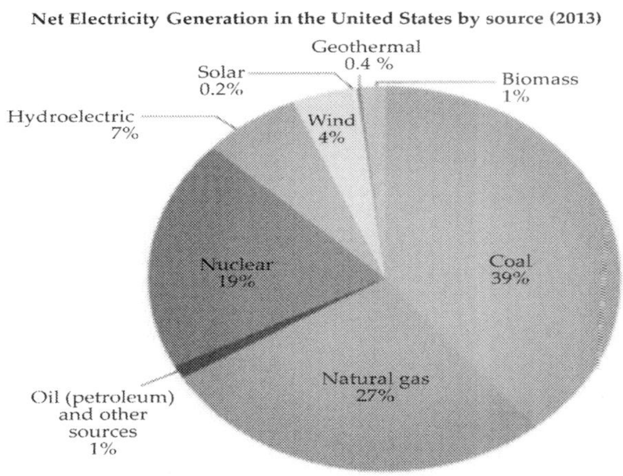

Figure 59: Net Electricity Generation in the United States by Source (2013).

Source: U.S. Energy Information Administration, (2014). Available at: https://www.eia.gov/electricity/data/browser/

Nearly all parts of the electricity system can affect the environment, and the level impact depends on how and where the electricity is generated and delivered (EPA, 2017). The impact may have a direct effect like the emission of air pollutants, or may have a passive effect like visual pollution or habitat modification. Even methods considered environmentally friendly, like wind, solar, and hydro, have some impact on the environment (Virginia Tech., 2007).

In general, the environmental effects of power generation can include (EPA, 2017):

- Emissions of greenhouse gases and other atmospheric pollutants, especially when a fuel is combusted;
- Use of water resources to produce steam, provide cooling, and serve other functions;
- Discharges of pollution into water bodies, including thermal pollution;
- Generation of solid waste, which may include toxic waste;
- Land use for fuel production, power generation, and transmission and distribution lines;
- Effects on plants, animals, and ecosystems that result from the air, water, waste, and land impacts above;
- Effects on human health, particularly if electricity generation results in humans being exposed to pollutants in air, water, or soil.

Virginia Technological University research (2007) argues that the whole cycle of electricity generation must be taking into account when looking at the environmental impact of power production. This cycle includes the production and transportation of fuel for the conversion process, as well as managing wastes or other pollutants. This is especially true for fossil fuel and nuclear power plants, which use large amounts of fuel mined from the earth. Energy sector environmental impact assessment process consists of fuel recovery and production, fuel transportation, electricity generation, electricity transmission, and spent fuel effects (Virginia Tech., 2007).

Environmental effects of raw fuel production:

Coal production. Coal is almost exclusively produced for electricity generation. Coal mining is often viewed as only a local environmental problem. Coal production, particularly surface mining, has both long-term and short-term impacts on landscapes and soils, including dust, noise, and water drainage/runoff. Preparation of coal produces both solid and liquid waste of which must be treated and disposed. Transportation of coal produces air pollution (i.e., dust), and coal storage produces dust to control and results in water runoff problems (Virginia Tech., 2007).

Oil production. The fuel oil that is used for combustion in power plants is a byproduct of the petroleum industry, so electricity production is partly responsible for environmental issues associated with oil and hydrocarbon burning emissions. Fuel burning process produces large volumes of greenhouse gases. Other environmental effects of oil production include blowouts, spills, brine disposal, and the production of hydrogen sulfide. Transportation of oil involves spill and leak risks. Oil refining can cause such environmental effects as explosions, fires, air emissions, noise, odor, and water runoff (Virginia Tech. 2007).

Natural gas production. During natural gas production, usual environmental impacts include blowouts, leaks, hydrocarbon emissions, and trace metal emissions. The treatment of natural gas involves hydrocarbons emissions into the air and the disposal of liquid residuals, while transportation and storage effects include spills and explosions (Virginia Tech. 2007).

Uranium production. Uranium mining involves such environmental issues as radioactive dust releases, mine water seepage, protection of workers from radioactivity, and the disposal of a large quantity of mine waste containing a low level of radioactivity. During treatment process mill tailings containing toxic metals and chemical wastes used in the treatment process, as well as radiological waste, are disposed in the environment.. Uranium enrichment process also results in radioactive dust releases, liquid and gaseous effluent releases, and must recycle fission products. Transportation of uranium involves the hazard of the accidental release of radioactive pollution (Virginia Tech. 2007).

Environmental effects of electricity production

Thermal power production: Fossil fuel power plants globally have the most extensive effect on the environment, as the combustion process produces air pollutants which spread over a wide area (Virginia Tech. 2007). The technological process of coal thermal power plant is shown at the figure 60.

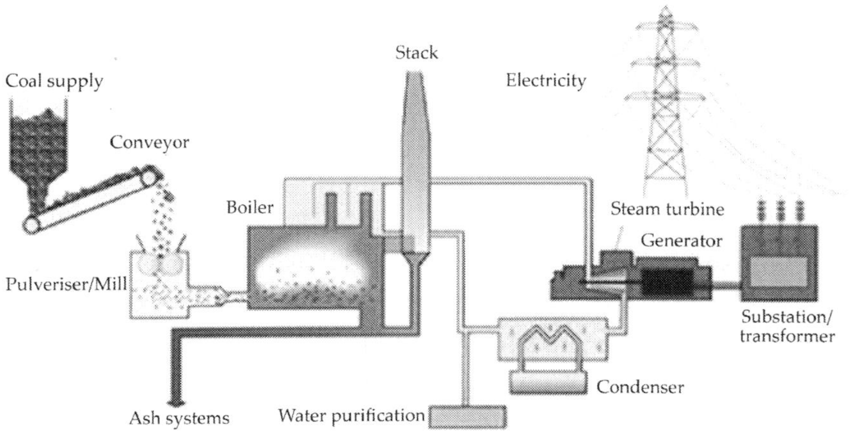

Figure 60: Electricity Generation at the Coal Thermal Power Plant.

Source: Bajajgroup, (2014). Available at: http://www.bajajgroup.org/blog/coal-thermal-power-plants-in-india

Thermal power plant works like generator: steam turbine moves the rotating machine that converts the mechanical energy into electrical energy that causes relative motion between conductor and magnetic field. Nuclear power plants also use thermal power cycle, but the role of thermal energy generator is played by nuclear reactor.

According to Virginia Technological University research (2007), fossil fuel power plants produce environmental problems including land and water use, air emissions, thermal releases, climatic and visual impacts from cooling towers, solid waste disposal, ash disposal (for coal), and noise. Thermal power plant needs large amounts of steam that results in significant effect on water use. For example, a typical 500 MW coal electrical generating plant uses 25 x 10^9 l/GW-year of water, which must be taken from a water source, and then cooled to return to the water source with as little environmental effect as possible (Virginia Tec, 2007).

The biggest effect fossil fuel plants have overall is the emission of air pollutants, particularly SO_X, NO_X, CO, CO_2, hydrocarbons, and water vapors. Carbon monoxide, CO, carbon dioxide, CO_2, nitrogen oxides, hydrocarbons, and even water vapor are the greenhouse gases, believed to be responsible for climate change. SO_X and NO_X are responsible for acid precipitations as they produce acid when released into the atmosphere (Virginia Tech. 2007). Table 2.6 list approximate amounts of air pollutants produced by combustion of different types of fuels.

Table 5: Thermal Power Plant Emissions (g/kWh)

Plant Type	**CO**	**NO_X**	**SO_2**	**CO_2**
Coal	0.11	3.54	9.26	1090
Oil	0.19	2.02	5.08	781
Gas	0.20	2.32	0.004	490

Source: Klymenko, (2000).

Nuclear power production: Nuclear energy harnesses the power of the atomic bomb in a slow, controlled and precise manner in order to gradually extract the wealth of energy available from splitting the atom (Kranzberg et.al., 2012). In a nuclear reactor, heavy water is superheated by the energy released from a series of blocks of radioactive material. This super-heated water is used to heat a tank of water which is not under pressure, which boils and spins a turbine that results in the generation of energy. The steam is then re-condensed, filtered back to the heating element, and the process begins anew (Peterson and Hermes, 2010).

Construction of a nuclear plant includes new roads, increasing traffic flow in the existing roads, excavations, cutting trees and other plants and impacts on ecology. Radioactive elements released from nuclear power plants are, however, the main object of monitoring and control to minimize exposure of the public to ionizing radiation. Thermal pollution from nuclear power plant impacts biota survival, growth and reproduction. Electricity production requires water consumption especially during the cooling process. The results of the water consumptions study by Shpant (2010) indicates that nuclear requirement for water consumption is extremely high due to the need for cooling of reactors, the highest of all the technologies within this study.

Shpant (2010) indicates that every day in the U.S, 2 to 2.5 gallons of water is used to produce a gallon of gasoline and with the aging of the wells water consumption will increase. Taking the highest range, nuclear withdraws 60,000 Gal/MWh more than wind. Fossil is second highest for withdrawal. There is a difference of 10,000 Gal/MWh between fossil and nuclear. At only 1g of water used per megawatt hour, wind has negligible water use. Photovoltaics also use very little water, approximately 30 gal/MWh. These low values for wind and photovoltaics indicate that they are more sustainable than nuclear and fossil fuel with respect to water consumption. There are major concerns about the treatment and risks associated with the generation and storage of radioactive wastes.

However, nuclear power plants have one environmental issue no other form of electrical power plant does. An accident at a nuclear power plant may release large amounts of radioactive particles, possibly resulting in a direct loss of life, and rendering a large land area immediately around the plant unlivable (Virginia Tech. 2007), as occurred at Chernobyl Ukraine in 1986, and most recently Japans Fukushima disaster in 2011.

The major regular environmental impact of normally operating nuclear power plant is the disposal of the high level nuclear waste contained in spent fuel rods, as this waste must be stored safely for thousands of years. There is no site in the United States currently accepting high level nuclear waste, so utilities are generally storing the waste in above ground casks at plant sites. A long-term issue is the decommissioning of nuclear power plants after the finishing exploitation life. At this point the whole reactor apparatus becomes a high level radioactive waste that must be disposed. Today's methodology of decommissioning suggests completely removing and disposing of entire plant, including all radioactive components, entombing the reactor in concrete. The other way is simply to shut the plant down and restrict access until the radioactivity will be neutralized (Virginia Tech, 2007).

Hydro power production: Electricity production using hydropower can have both positive and negative effects on the environment. In some areas, a dam may help with flood control, flow regulation, or the reservoir may provide recreational opportunities. At other territories, the dam may have adverse effects on the hydrological cycle, water quality of the stream, stream ecology, fish migration, and cause the damaging of landscapes and ecosystems. Building the new high-head dams requires local population resettlement and compensation costs. Low-head dams usually have a benign effect on the environment. During the heavy rains dam can be ruined by overflow that leads to catastrophic floods (Virginia Tech, 2007).

Environmental effects of electricity transmission

Transmission of electricity from the generating plant to the consumers uses wires suspended on large towers, known as transmission lines. Traditionally these lines have been viewed only as an aesthetic nuisance that could cause communications interference and be a hazard to low flying aircraft (Virginia Tech. 2007). Today, according to Virginia Technological University research (2007), there are other issues considered about the effect of transmission lines on the environment. Greater concern is placed on the effect of the lines on the natural habitat. The main new problem is the effect of electromagnetic fields (EMFs) on human health.

More than 1000 studies have been performed since 1979 to evaluate the connection between low frequency magnetic fields and people health. A lot of research focused on the relationship between cancer rates and fields produced in the 50-60 Hertz range used for electricity transmission. The studies generally were concentrated on children with residential exposure to high voltage transmission lines, or workers with high degrees of occupational exposures to EMFs. Most of the studies failed to establish a statistically significant relationship between cancer and electromagnetic fields. One issue in the studies was to estimate the amount of EMF exposure a person receives (Virginia Tech, 2007).

Magnetic fields are produced when electrical current flows through a conductor. The strength of the magnetic field drops quadratically with distance from the conductor. This means that while the strength of a magnetic field directly underneath a transmission line is in the range of 300 to 600 milliGauss (mG), the field strength is only 10 to 100 mG 200 feet away. To compare, the earth's natural magnetic field is approximately 450 mG. Home electric tools like hair dryers, microwaves, and televisions produce magnetic fields, ranging in strength from 1 to 50 mG. Obviously, due to the large number of different sources of EMFs, estimating a person's daily exposure becomes extremely complex. The studies generally used a broad estimate of EMF exposure (Virginia Tech, 2007).

Environmental impacts of distributed generation

Distributed generation can benefit the environment as its use reduces the amount of electricity that must be generated at centralized power plants, in turn can reduce the environmental impacts of centralized generation. Specifically:

- Existing cost-effective distributed generation technologies can be used to generate electricity at homes and businesses using renewable energy resources such as solar and wind.
- Distributed generation can harness energy that might otherwise be wasted—for example, through a combined heat and power system.
- By using local energy sources, distributed generation reduces or eliminates the "line loss" (wasted energy) that happens during transmission and distribution in the electricity delivery system (EPA, 2017).

However, according to US Environmental Protection Agency (2017), distributed generation can also lead to negative environmental impacts:

- Distributed generation systems require a "footprint" (they take space), and because they are located closer to the end-user, some distributed generation systems might be unpleasant to the eye or cause land-use concerns.
- Distributed generation technologies that involve combustion—particularly burning fossil fuels—can produce many of the same types of impacts as larger fossil-fuel-fired power plants, such as air pollution. These impacts may be smaller in scale than the impacts from a large power plant, but may also be closer to populated areas.
- Some distributed generation technologies, such as waste incineration, biomass combustion, and combined heat and power, may require water for steam generation or cooling.
- Distributed generation systems that use combustion may be less efficient than centralized power plants due to efficiencies of scale.

- Distributed energy technologies may cause some negative environmental issues at the end of their useful life when they are replaced or removed.

As we can see, energy sector today is responsible for a significant impact on environment and human health. Using key concepts and tools of Industrial Ecology, humans can design more sustainable energy generating sector. US EPA proposes such solutions to reduce the negative environmental impacts associated with generating electricity:

- Energy efficiency. Energy efficiency is a resource that reduces the amount of generated electricity and thus, level of environmental pollution. In this respect, end-users can meet some of their needs by adopting energy-efficient technologies and practices. (EPA, 2017).
- Clean centralized generation. New and existing power plants can reduce environmental impacts by increasing generation efficiency, implementing the pollution control systems, and leveraging cleaner energy supply resources (EPA, 2017).
- Clean distributed generation. Distributed generation, especially distributed renewable energy, can help support delivery of environment friendly, reliable power to customers and reduce electricity losses along transmission and distribution lines.
- Combined heat and power (CHP) production. Also known as cogeneration, combined heat and power, produces electricity and heat simultaneously from the same fuel source. By using heat that would otherwise be wasted, CHP is both distributed generation and a form of energy efficiency (EPA, 2017).

Renewable energy technologies can be helpful in achieving the inclusive and sustainable development of energy sector. Renewables have experienced rapid deployment over the past few years, mainly driven by the ambition to improve energy security, enhance energy access, and mitigate climate change. Many countries are now exploring ways to stimulate social and economic growth through the development of the renewable energy sector. Wind and solar photovoltaic (PV) are currently the fastest-growing sources of electricity globally. According to the International Energy Agency (2016), in 2015, their additional annual generation met more than 90% of incremental demand for electricity.

Between 2008 and 2015, the typical cost of land-based wind turbine decreased by 35% and that of solar PV decreased by almost 80% (IEA, 2016).

Today, increasing evidence of a new phase of deployment can be encapsulated by the notion of next-generation wind and solar power; in this phase, wind and solar PV are technologically mature and economically affordable. Technological maturity and lower costs make wind and solar power an increasingly attractive option for policy makers seeking to meet energy policy objectives, such as improving energy security by diversifying supply, reducing local pollution and reducing CO_2 emissions. Wind and solar power are expected to make a significant contribution to meeting the objectives of the Paris Agreement. Their contribution to electric systems around the globe is rapidly moving from alternative to mainstream, including in emerging and developing countries (IEA, 2016).

The success of renewables is driving change in power systems around the globe. Electricity generation from both technologies is constrained by the varying availability of wind and sunshine, which makes their output variable over time. Where flexible renewable energy is growing beyond a few percentage points as a share of annual electricity generation, power system operation and planning are being upgraded and adapted to accommodate variable renewable energy on the grid. These adaptations generally ensure a reliable balance between electricity supply and consumption, even in the presence of double-digit shares of renewables in annual generation. For example, coal electricity generating plants in Denmark have been upgraded to allow for more flexible operation, starting plants quickly and allowing them to change their output across a wide range. In Germany, the four transmission system operators began sharing certain reserve capacities in 2008, which has allowed to reduce their combined reserve requirements despite ongoing renewable energy deployment (IEA, 2014).

Electricity generation with solar energy sources generally has less environmental impact. There are no emissions and sewage water produced in the energy conversion process. The only exception is solar thermal collectors, which have an operating liquid that must occasionally be discharged. Some other environmental concerns include need in large land area for PV stations location and thermal pollution as solar plants produce a great deal of heat. An unknown quantity in solar energy is the disposal of photovoltaic cells. The most promising solar cells use gallium arsenide, a toxic substance (Virginia Tech, 2007).

The major environmental effects of wind power plants come from visual pollution, noise, and TV interference. This is mainly true for wind farms, where 50 or more wind turbines may be mounted at the same site. If wind farm was built on a bird migratory path, it may pose major hazards to birds. The occupational safety hazard is a possibility of blade breakage (Virginia Tech, 2207).

International Energy Agency (2016) admits that the greater uptake of distributed renewables shifts the historic balance of supply and demand in the power network, and calls for a revision of the institutional rules guiding low- and medium-voltage grids. The rise of distributed generation assets – dominated by the rapid uptake of solar photovoltaic – translates into growing complexity of power flows within the power distribution grid (Figure 61). This provokes the demand for innovative approaches to the planning and operation of low- and medium-voltage grids, with technical, economic, and administrative implications (IEA, 2016).

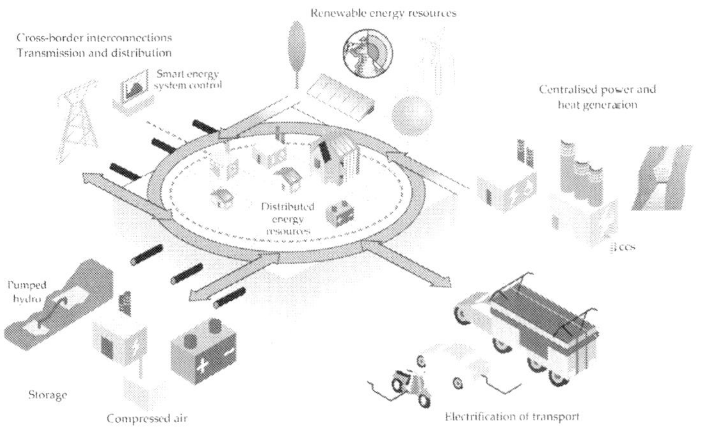

Figure 61: Smart Distribution Grid.

Source: OECD/IEA, (2016). Available at: https://www.iea.org/publications/freepublications/publication/NextGenerationWindandSolarPower.pdf

Power system transformation implies a change in basic assumptions of low- and medium-voltage grids – away from passive distribution of electricity and towards becoming a critical hub for electricity and data. On the technical side, more dynamic and bi-directional flows of electricity (from lower to higher voltage levels and vice versa) require reinforced monitoring and control capabilities as well as upgrades to infrastructure. Moreover, designing standards need upgrading to manage the uptake of large shares of distributed resources. In this context, next-generation variable renewable energy technology – such as advanced inverters – can offer technical abilities to support and sustain safe and reliable functioning of local power grids, while also reducing energy losses in the overall power system (IEA, 2016). In residence sector, transition to renewable energy sources - wind, solar, geothermal, biomass - can, on the one hand, provide consumers with flexible local power plants, and, on the other hand, what is the most important - it can significantly reduce pollution. Implementation of wind and solar energy in heat-and-cool-supplying systems of buildings is an effective instrument in building sustainable society. Figure 62 presents sustainable energy system for residential building based on balanced use of traditional (natural gas and electricity) and renewable (solar and wind energy) sources.

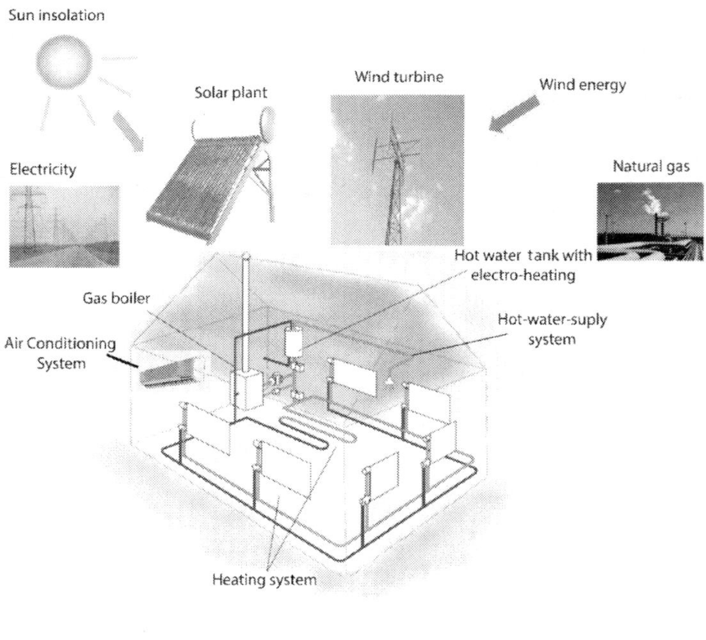

Figure 62: Sustainable System of Heat-and-Cool Supplying for Residential Building.

Improving of environmental safety of heat-and-cool-supplying needs to develop the algorithm of optimal redistribution of power between renewable and traditional sources between heating, cooling, and hot water systems. The algorithm of energy redistribution within the alternative system based on wind and solar energy with compensation of energy shortage by conventional sources is shown at Figure 63.

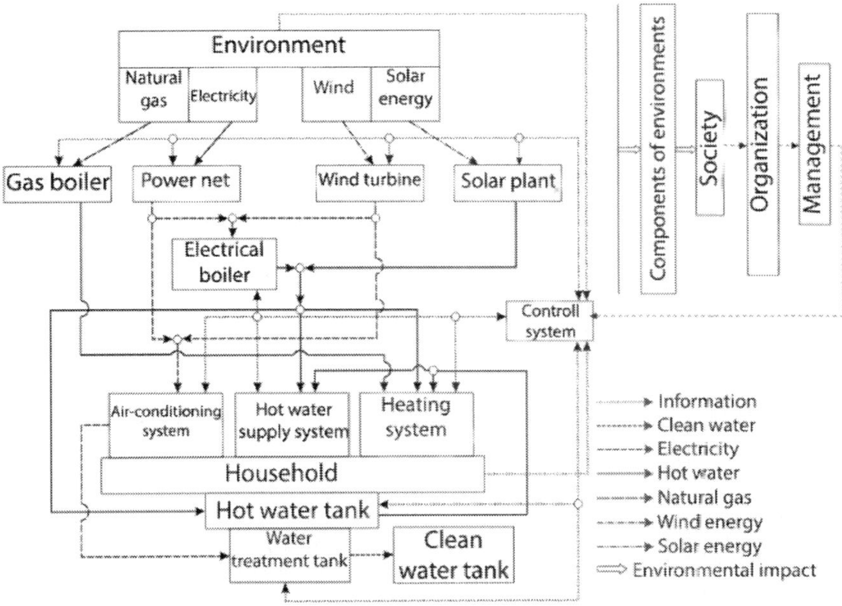

Figure 63: Algorithm of Increasing the Environmental Safety of Heat-and-Cool-Supplying by Sustainable Use of Renewable and Traditional Energy Sources.

Proposed decision is based on redistribution of wind and solar energy between hot-water-supplying, heating, cooling, and water treatment with a maximum level of substitution of traditional resources by renewables. Simulation results show that accounting of energy flows redistribution provides more efficient use of renewables by 15-20%; and the use of optimal solution makes it possible to increase the sustainability level alternative system compared with traditional coal-burning by 63%. Using of renewable energy in systems of heat-and-cold-supply will reduce air pollution by 43-66% and the volume of generated solid waste by 46-62% compared with traditional system based on electricity produced by coal-generating plant (as the worst option).

METALLURGY

Metallurgy (Metal Industry) is the industry of metals: the way in which science is applied to the production of metals, and the engineering of metal components for usage in products for consumers and manufacturers.

Metal industry is subdivided into ferrous metallurgy (or black metallurgy) and non-ferrous metallurgy (or colored metallurgy). Ferrous metallurgy involves processes and alloys based on iron, while non-ferrous metallurgy involves processes and alloys based on other metals. The production of ferrous metals accounts for 95 percent of world metal production (Klymenko, 2000).

Metallurgical activities include ore extraction and processing, melting processes, development of alloys and casting in different forms depending on the request. Generally, extraction of the ore, flotation and training material to storage dumps is an modification of the natural environment involving changes in ecosystems and dynamic balance that have serious consequences on their durability (Sehic-Music et.al., 2013).

The cast obtaining process causes negative environmental effects through the such technological stages: preparation core training, development, casting and ingots, pieces debate, parts cleaning, and reshuffling. Melting induce soil, vegetation, air, and water contamination with metals (Habashi, 2011). Training materials after use are usually dumped in the disposal that often are not waterproofed, and are sources of pollution to groundwater. Sometimes such industrial disposals are improperly designed and become a reason of lateral leaks which induce soil and vegetation pollution in the surrounding area. Wind blows out fine particles from waste dumps and disseminate them over larger surface producing pollution that harms plants and enter in the food chain. Physical pollution of soil is caused both by stockpiling waste and through forms that are no longer used and are not melted (Sehic-Music et.al., 2013).

The environmental impact of metallurgical industry causes a deterioration of natural ecosystems and produces imbalances in natural cycles with serious consequences on their sustainability. The main sources of environmental impact of metallurgical activities are (Cirtina et.al., 2016):

- ferrous industry that generates pollutants such as mineral compounds, coal, ash, cyanides, phenols, acids waters;
- non-ferrous industry through activities for obtain copper, zinc, aluminum generators cyanides, acids, heavy minerals, chlorine.

Foundry industry causes pollution by excavation works (mining, gravel pits and quarries); pollution by covering the soil with dumps, tailings ponds, waste facilities; pollution with sewage and inorganic waste, metals, salts, basic acids; pollution with airborne substances. Soil is polluted mainly due to waste and second products; especially harmful is the dross waste as it kills soil microflora, damages the vegetation and can even cause technogenic desertification of the territories where waste was disposed (Cirtina et.al., 2016).

Steel industry wastewater contains the entire range of organic and inorganic waste products, specific for production units (Cirtina et.al., 2016). Table 6 represents the sources of pollution and pollutants in wastewater from the steelworks.

Table 6: Sources of Water Pollution from Steel Industry

Section name	Wastewater pollutant
Furnace coke plant	Heat, ammonia, phenols, cyanides
Furnace	Heat, suspended solids, cyanide, sulfur compounds
Steelworks	Heat, suspended solids, lime, oil, dust, iron carbonates and bicarbonates dissolving of fluxes used in smelting
Hot rolling	Heat oil, iron oxide and petroleum products
Cold rolling	Oil emulsion, heat
Finishing Processes	Alkali, emulsions, acids, iron salts, phenols, fats, solids

Source: Cirtina et.al., 2016. Available at: http://hrcak.srce.hr/file/226425

There are two categories of water discharged from coke-chemical plants, according to their origin and content of contaminants:

- phenol ammonia water containing phenols, ammonia, tars, oils, organic compounds;
- phenolic waters contaminated with phenols, ammonia, benzene, sulfide hydrogen, cyanide, tars, oils, organic compounds (Cirtina et.al., 2016).

The range of pollutants present in these two categories of sewage waters highlights the risk they pose to the natural receptor. Technological processes which contribute to getting parts through metal casting, release a large amount and variety of pollutants into the air, water, and soil (Cirtina et.al., 2016):

- dust – dust particles that form the dispersed segment and air gas that forms air dispersion medium segment. Powders sizes between 0,01 - 10 microns;
- the smoke resulting from the melting of metal and metal oxide consist of solid particles sized less than 1 micron;
- fog formed of fine liquid particulates.

The dust produced by the steel works can be in quantities of up to 80 % of iron oxides (red powder), and optionally manganese, and the iron and steel foundries containing oxides of silicon, aluminum, calcium, zinc, etc. (Cirtina et.al., 2016). A particular aspect of these powders could be the high absorbency by water and feather reactions with SO_x and NOx, thereby increasing their adverse effects on human health.

Let us analyze the environmental impact assessment of metalwork site presented in the research of Cirtina et.al (2016). Environmental Impact Assessment was proceeded using a complex matrix method based on cause-effect relationship (Rojanschi Matrix) and the calculation of the entire system pollution. In the method, each medium separately, each type of source and each pollutant fall on a scale of creditworthiness expressed through classes from 1 to 10, where 10 is the natural state unaffected by industrial activity and 1 is a situation with irreversible damage of the environmental factor detected (Cirtina et.al., 2016).

For the evaluation of the global pollution index (GPI), calculated marks for each component of the environment, are transferred on a scale of separate creditworthiness, which is divided into 6 classes with values between 1 and 6 in which: class 1 – is the natural environment unaffected by human activity and class 6 – is degraded environment, improper forms of life. Figure 64 shows us the matrix of global pollution index assessmen for the environmental components in the site area of non-ferrous metal foundries (Cirtina et.al., 2016).

Environ-mental Compo-nents	Cause: Pollutants emissions Effects: Notes on creditworthiness scale						NC
Air	SO_2						8
	NO_2						7,5
	CO						8
	powders						7,5
	VOCs						8
	aldehy-des						8
	Metals						7
Water		pH					8
		Ether extractable oil					7,5
		Materials in suspen - sion					7,5
		BODs					8
		COD					8
		Deter-gents					8
		Ammonia nitrogen					8
		Sulphi-des					8
Soil-vegeta-tion			SO_4^{2-}				9
			N				9
			Metals				9
			VOCs				9
Noise							8
Waste							7,5
Popula-tion							8
Total	7,8	7,9	8,5	8	7,5	8	7,84

Figure 64: Matrix Assessment of Global Pollution Index in the Metal Work Site.

Source: Cirtina et.al., 2016. Available at: http://hrcak.srce.hr/file/226425

The rating scale for the global pollution index represents the environmental impact or effect of human activities on the environment.

The pollution index value was calculated for environmental components: air, water, soil, vegetation, noise, waste, human health status. Global pollution index was simulated GPI = 1,879. According to the scale of the impact assessment, the value GPI is a Class 1 - 2 and indicates an environment subjected to industrial impacts within acceptable limits (Cirtina et.al., 2016).

Some scientists refer conventional steelmaking as one of the world's leading industrial sources of greenhouse gases (Klymenko, 2000). Massachusetts Technological University research (2013) developed a new process that could change all that—and produce stronger (and ultimately cheaper) steel. As MIT informs, worldwide steel production currently totals about 1.5 billion tons per year, and each ton produced generates almost two tons of carbon dioxide, according to industry data. This accounts for about 5 percent of the world's greenhouse-gas emissions. The idea for the new method, which was developed by materials chemistry professor Donald Sadoway, assistant professor of metallurgy Antoine Allanore, and Lan Yin, PhD, arose when Sadoway received a grant from NASA to look for ways of producing oxygen on the moon—a key step toward future lunar bases. He found that a process he invented called molten oxide electrolysis could use iron oxide from the lunar soil to make oxygen in abundance (Chandler, 2013).

This method used an iridium anode, but since iridium is expensive and supplies are limited, that's not a viable approach for bulk steel production on Earth. Finding an alternative wasn't easy, because molten iron oxide, at about 1,600 °C, "is a really challenging environment," Sadoway explains. "The melt is extremely aggressive. Oxygen is quick to attack the metal". The answer was an alloy that naturally forms a thin film of metallic oxide on its surface—thick enough to prevent further attack by oxygen but thin enough for electric current to flow freely through it. The alloy's constituents – iron and chromium – are abundant and cheap (Chandler, 2013).

Professor Sadoway and his team admit that in addition to producing no emissions other than pure oxygen, the process lends itself to smaller-scale factories. Conventional steel plants are profitable only if they can produce millions of tons of steel per year, but this new process could be viable for production of a few hundred thousand tons per year, he says. The process also yields metal of exceptional purity, Sadoway says. And it could be adapted for carbon-free production of other metals and alloys, including nickel, titanium, and ferromanganese. (Chandler, 2013)

The technology is still at the laboratory scale, but MIT team has formed a company to develop a commercially viable prototype plant. Designing, building, and testing such a facility could take about three years and after that technology will be ready to broad implementation.

The other example of environmental friendly technology in metal industry is powder metallurgy (PM). Addressing manufacturing processes, powder metallurgy's sustainable value is primarily derived from its net-shape capabilities and its very high material-utilization factor, which minimizes all energy inputs. As Powder Metallurgy Company admits (2012), in general, any metal component can be manufactured by any of several manufacturing technologies. A simple gear can be made by machining a cylindrical piece of solid bar stock, forging a steel blank in forging dies, in some cases stamping it from sheet or roll stock, possibly casting it and after this machining features, or in the case of PM compacting powder in tooling dies that result in the product's final shape. Evaluating the sustainability of a product's manufacture, we usually compare the process steps, resources, and economic costs that go into the manufacture of that product. Producing a component by the powder metallurgy techniques, such as the gear shape in the Figure 65, we can realize that there is no waste in the shaping of the component.

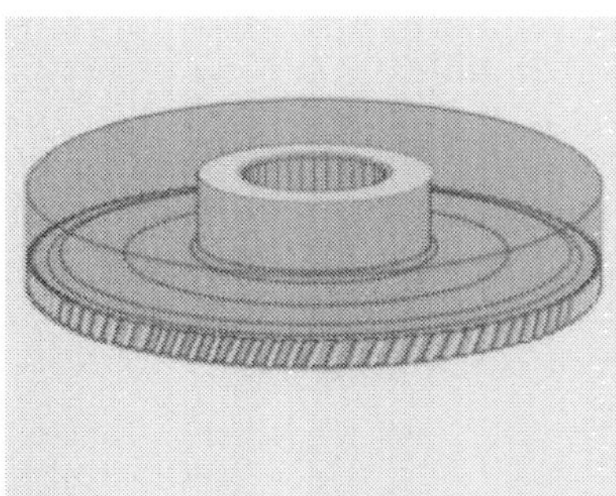

Figure 65: Gear Shape.

Source: Powder Metallurgy Company. (2012). Powder Metallurgy—Intrinsically Sustainable. Available at: http://powdermetallurgyco.com/a-green-technology

Production of the gear by chip-generating machining would result in 40% waste as material would be cut away and discarded. While the machining waste can be recycled, it is extraneous to the final component and is a net loss to the material and energy eco-efficiency of the final product. Studies and practice show that 85% of all PM powders are produced from recycled material. Most metals can be repetitively recycled in significant quantities. The major amount of metal powder used, (iron/steel), close to half million tons per year, is nearly always produced via atomization of electrically melted steel scrap. Particle-size distribution in each sample is controlled by sieving operations, and if the element size needs to be adjusted, the powders can be milled to avoid waste of oversized particles (PMC, 2012).

The net-shape nature of powder metallurgy decreases the energy demand per component. All metallurgical processes require the use of thermal, chemical, or mechanical energy to achieve product form. Some processes require several heating and re-heating operations to achieve final form. The only time metal for powders is melted is in the atomization step; all other thermal operations are undertaken below melting temperature, conserving energy while achieving the final shape and developing the necessary material properties/mechanical performance. In addition, finishing to final product specifications usually doesn't need to be done, further conserving the energy necessary to achieve final product characteristics. The chart below (Figure 66) compares the energy consumption per kilogram of completed components for the various processes; only casting and deep-drawing operations are in the general range of PM's energy consumption (PMC, 2012).

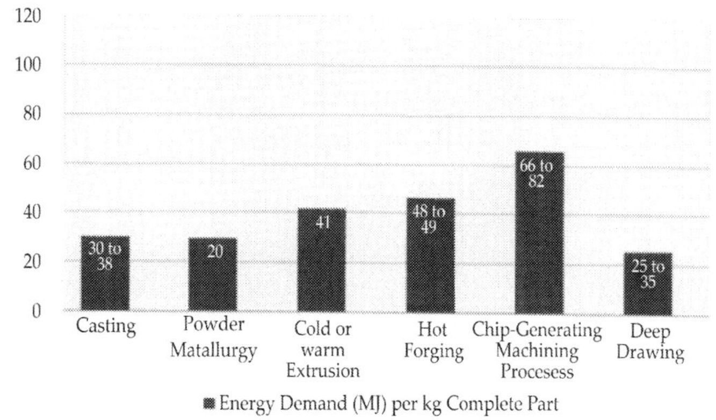

Figure 66: Energy Consumption of Completed Components for the Various Metallurgical Processes.

Source: Powder Metallurgy Company. (2012). Powder Metallurgy—Intrinsically Sustainable. Available at: http://powdermetallurgyco.com/a-green-technology

As we can see from Figure 66, the comparative analysis of per-piece energy consumption demonstrates 44% less energy is required to produce the PM version than the forged and machined part.

Talking about environmental effects, we must say, that powder manufactured components can be given a wide assortment of secondary finishing operations such as machining to exacting tolerances, plating, or coating. According to Powder Metallurgy Company. (2012), on a per-kilogram of-finished-component basis, PM final machining operations are minimal, in most cases resulting in minimal use of cutting oils per pound produced. Also, most plants only emit cooling water to public water/septic systems, minimizing the likelihood of toxic releases due to these point-source contaminants (PMC, 2012).

The PM process does require the use of graphite or other lubricants in the compaction stage, but recent improvements in both binder and lubricant technologies has resulted in most companies shifting from chemical systems using listed constituents such as zinc stearates to more innocuous materials in order to minimize or eliminate the potential for gaseous toxic releases from sintering operations. Gaseous emissions are also minimized by the net-shape conservation of product mass affecting releases. With increased automation, more customers are deploying returnable shipping containers and crates in an effort to shrink the volume of cardboard that finds its way to landfills. Since nearly all scrap produced is metallic, it is routinely recycled, thus minimizing the industry's contribution to landfills. Compared with other manufacturing processes, and considering the low energy intensity per pound of product minimizes overall environmental impact, powder metallurgy displays less environmental hazards (PMC, 2012).

MECHANICAL ENGINEERING

Mechanical Engineering or Machinery Industry is a subsector of the industry, that produces and maintains machines for consumers, the industry, and most other companies in the economy. Machine industry traditionally belongs to the heavy industry.

International Trade Administration (2016) informs that machinery manufacturing is one of the largest and most competitive sectors of the U.S. manufacturing economy. After a sharp decline in exports in 2009 and subsequent recovery, exports of capital equipment totaled $166 billion in 2015. Among the wide range of machinery manufactured in the United States, construction machinery, engine equipment, turbines and turbine generator sets, industrial process controls, and agricultural equipment led U.S. exports in 2015. Major competitors in global machinery markets include Germany, Japan, Italy, and China (SelectUSA, 2016).

Machinery industries provide essential and highly sophisticated technology for many other manufacturing and service industries. Industrial process controls and other automation technologies enable end-users to maximize the productivity of their equipment. Sales of many types of machinery are accompanied by a variety of high-value services as well, including specialized architecture, engineering, and logistics.

According to the International Trade Administration, there are such Mechanical Engineering Industry Subsectors:

- Agricultural and Food Machinery and Equipment: Equipment used to grow, process, package, transport, and distribute food and beverages. Agricultural and Food Machinery and Equipment includes agricultural equipment, food processing and packaging machinery, commercial and industrial refrigeration equipment, and commercial food service equipment.
- Aerospace: Machinery in this subsector include machine tools, material handling equipment, welding equipment, and process control technology.
- Automotive: Automotive machinery and related systems, including machine tools, material handling equipment, welding equipment, and process control technology make up this subsector.

- Construction: Construction and mining equipment and related systems. This subsector includes self-propelled equipment, implements, accessories and components for use in construction, forestry, mining, and utilities. Off-road diesel engines and fluid power technology are also important components of off-road equipment.
- Energy: Equipment and machinery for generating, transmitting and distributing electric power as well as machinery used for oil and gas exploration and production. Major categories include turbines, power transmission equipment, and internal combustion engines (except automotive gasoline and aircraft) for electric utility and industrial applications. Also includes oil and gas field machinery.
- Mining: Equipment and machinery used to mine and extract natural resources. These include mining and materials handling equipment and process controls, pumps, electric motors, and fluid power (SelectUSA, 2016).

Machine building process consists of four main stages:

- Workpieces production:
- Machining (finishing);
- Assembling;
- Testing.

Most of the activities related to the machine industry, directly or indirectly, have significant impact on the environment and human health. Direct activities are connected to the production and use of the machines, while indirect are related to all other activities that are not controlled by the sector but influence its overall picture such as the production and distribution of raw materials (Poulikidou, 2010). Figure 67 presents schematic analysis of environmental impact of automobile production.

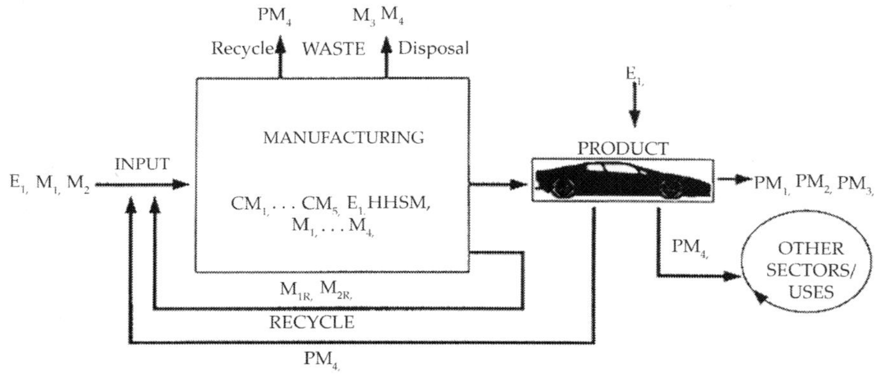

Figure 67: Environmental Performance Metrics in Automobile Production.

NOTE: CM[1] = Manufacturing emissions (other than TRI, 33/50, GLPT, and SARA Title II chemicals); CM_2 = Toxic Release Inventory chemicals; CM_3 = 33/50 chemicals; CM_4 = Great Lakes persistent toxics; CM_5 = SARA Title III chemicals; E_1 = Energy; $M_1 + M_{1R}$ = Parts, components, raw materials, and recycled materials; $M_2 + M_{2R}$ = Water and recycled water; M_3 = Packaging waste; M_4 = Solid waste (excluding packaging); PM_1 = Tailpipe HC, NO_x and CO emissions; PM_2 = Evaporative emissions; PM_3 = Tailpipe CO_2emissions; PM_4 = Recycled material from manufacturing process and product; HHSM = Human health and safety.

Source: National Academy of Sciences. (2017). The Automotive Industry. In: Industrial Environmental Performance Metrics Challenges and Opportunities, ISBN 0-309-06242-X, 56-63. Available at: https://www.nap.edu/read/9458/chapter/6#82

Talking about environmental effects of machine manufacturing, we should take in account the acquisition of the required resources for materials and energy productions that begins with mining. Mining processes based on surface and underground technologies are used for the extraction of coal, iron ore, bauxite ore etc. Raw materials extraction usually has not only local, but also global effects. Some of the most important local effects of mining stage are: landscape degradation, deterioration and pollution of areas used for vegetation, deforestation, occupation of large excavation areas, changes in land use, changes to hydrological and geological characteristics of the local area etc.

(Keoleian, 1997). At a more global level, mining can result in extinction of plant and animal species contributing to the irreversible deterioration of the biodiversity (National Academy of Sciences, 2017).

The same environmental effects are caused from finishing machinery processes, especially drilling processes, which are necessary in order to derive the oil that is needed for almost all activities connected to the machine production (Keoleian, 1997). Land occupation is also connected to landfill as a method of waste final disposal. According to the National Academy of Sciences (2017), waste is produced in every stage of the machine's life cycle and there is a considerable flow of materials that are disposed to landfills.

The main consumables used in a machine tool during the workpiece processing are (Klymenko, 2000):

- Cutting tool;
- Cutting fluid;
- Lubricant oil;
- Filter.

During the production phase, almost 70% of the overall environmental impact is caused by the electricity consumption. In order to avoid or decrease effects of this situation, energy efficiency technologies for the machine tools should be implemented (Gonzalez, 2007).

Water is one of the important resources in machine building complex. A high percentage of the cutting fluid is composed of water, which is mixed with oil and chemicals (Gonzalez, 2007). There are different cutting fluids (Klymenko, 2000):

- Oil based cutting fluids:
    - Straight oils;
    - Emulsions-soluble oils.
- Chemical generated cutting fluids:
    - Synthetics;
    - Semi synthetics.

Jeffrey B. Dahmus and Timothy G. Gutowski study (2007) states that 4.07 gallons/machine/day of water is consumed for cutting fluids, which is 15.38 liters for every machine per day. Impacts in the water stream associated with cutting fluid disposal are (Gonzalez, 2007):

- Hazardous chemical compounds;
- Hazardous metal carry-off (workpiece material);
- BOD;
- Nutrient loading;
- FOG.

Spent cutting fluid can become hazardous when EP additives, specially chlorine compounds (most common – chlorinated paraffin), are added. In addition, if working with metals other than carbon steel, there is a possibility that heavy metals (cadmium, copper, chromium, lead, mercury, nickel, silver, zinc) can be taken by the fluid and cause water pollution (Gonzalez, 2007).

Cutting fluids usage can cause health problems. According to the US Environmental Protection Agency (1995), workers in machining operations are continually exposed to cutting fluids. A fluid must be relatively non-toxic, non-flammable and non-misting to decrease environmental and health risks. Dermatitis and respiratory problems are the most frequent health effects of cutting fluids usage. Dermatitis usually is the result of direct contact with the fluid, whereas respiratory problems are caused by oil mist generated in the work environment (Gonzalez, 2007). Figure 68 presents the example of air pollutants distribution among the different life cycle stages of automobile.

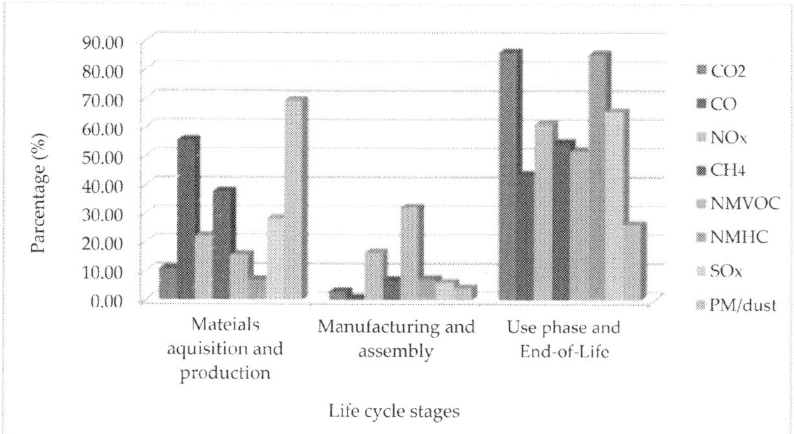

Figure 68: Redistribution of Emission of Substances in the Air Among the Different Life Cycle Stages of an Automobile.

NOTE: (CO2=carbon dioxide, CO= carbon monoxide, NOx= nitrogen oxides, CH4= methane, NMVOC=non-methane volatile organic compounds, NMHC= non-methane hydrocarbons, SOx= sulfur oxides, PM= particulate matter).

Source: . Identification of the main environmental challenges in a sustainability perspective for the automobile industry. Available at: http://publications.lib.chalmers.se/records/fulltext/136380.pdf

Guardian News (2017) calculated the carbon footprint of a new car:

- 6 tons CO_2e: Citroen C1, basic spec;
- 17 tons CO_2e: Ford Mondeo, medium spec;
- 35 tons CO_2e: Land Rover Discovery, top of the range.

During the manufacturing phase, different waste types are generated (Klymenko, 2000):

- Spent filters.
- Spent Coolant.
- Sludge.
- Spent lubricant oil.
- Metal chips.
- Spent cutting tools.

Whereas some of these by-products can be disposed at the landfill, other must be treated as special or hazardous wastes. This distribution highly depends on cooling fluids' mixture. Metal chips usually are separated from cutting fluids and can be recycled as scrap (Gonzalez, 2007).

As mentioned before, environmental factors play an important role in designing and planning the manufacturing processes, and could become as important as cost factors for engineers. Energy distribution and efficiency has a significant impact, not only on cost but also on environment. The Life Cycle Analysis concept of Industrial Ecology can assist the designer to develop a more environmentally friendly product (Gonzalez, 2007).

However, the machine building industry sector made progress in environmental technology. Therefore, there are currently many research studies that are working on the machine tool improvement, not only in performance areas but also in more environmental efficient and safe machining (Gonzalez, 2007).

The integrated project NEXT (Next generation production systems) aims at committing the main players at European machine building industry market to new frontiers in diverse fields (NEXT, 2011). The main objectives of NEXT project relate to:

The green machines: Innovation forward in environmentally friendly production machines. The aim is to get machines that consider environmental aspects through their entire life-cycle:

- Use of recycled materials for machine elements (>50%);
- Reduction of energy consumption (>25%) at machine use;
- Zero waste produced;
- Pull to pieces and recycling 100% machines;
- Non-pollution alternative processes;
- Reduction of machine volume;
- Reduction of noise pollution;
- Reduction of logistic area for maintenance/spare parts allocation.

The user centric autonomous machine: Breakthrough in usability and machine autonomy;

The manufacturing breakthrough: Big leap in performance and innovative processing methods;

New business pattern for machinery: A breakthrough to create value for the production;

New contents for training, marketing, socio-economics aspects, and dissemination around production (NEXT, 2011).

The other example is the study "CENIT of Machine tools" dealing with eco-machines or e-machines. Its main goals are:

Reduction of the weight of the structural elements using lighter materials. Currently machine tools are basically made of cast iron, welded steel, and a several aluminum alloys. There is already a cluster working on this called 'Lightstruct' (Light Structures Machine Tools). In addition, materials used in machine tools should also increase amortization properties to absorb the impacts.

Use of ecologic lubricants and refrigerants, which will be biodegradable and nontoxic.

Energy optimization:

- Focusing on auxiliary systems, such as hydraulic systems, pumps. These systems are usually working even though the main machine is switched of or when these systems are not needed. An efficient management of these auxiliary machinery could save 10% of the total consumption.
- Energy demand management:
- Energy recovery (CENIT de MH, 2010).

CHEMICAL INDUSTRY

According to the Organization for Economic Co-operation and Development (2001), Chemical Industry is very diverse, comprising basic or commodity chemicals; specialty chemicals derived from basic chemicals (adhesives and sealants, catalysts, coatings, electronic chemicals, plastic additives, etc.); products derived from life sciences (pharmaceuticals, pesticides and products of modern biotechnology); and consumer care products (soap, detergents, bleaches, hair and skin care products, fragrances, etc.). Chemicals are used to make virtually every man-made product and play an important role in the everyday life of people around the world. Such products provide protection for crops and increase yields, prevent and cure disease, provide insulation to reduce energy use and provide countless other benefits that make life better for people.

US Chemical Manufacturers Association (CMA) and US Environmental Protection Agency define Chemical industry as the industry that includes such sectors:

- Chemicals and allied products;
- Industrial organic chemicals;
- Plastics materials and synthetics;
- Pharmaceuticals;
- Soaps, cleaners, and toilet goods;
- Paints and allied products;

- Industrial organic chemicals;
- Agricultural chemicals (incl. fertilizers and pesticides);
- Miscellaneous chemical products.

Considering previous definition, Industry and Analysis Unit of the US International Trade Administration points out such subsectors within the chemical industry (SelectUSA, 2016):

Basic chemicals: organic and inorganic chemicals, plastic resins, dyes, and pigments. Plastic resins, in particular, have experienced significant growth as a replacement for traditional materials in the automotive, construction, and packaging end-use markets.

Specialty chemicals: adhesives and sealants, water treatment chemicals, plastic additives, catalysts, and coatings. These chemicals are performance-oriented and typically include customer/technical servicing as an aspect of their sales.

Agricultural chemicals: play a crucial role in the farm economy and the food processing sector. Thanks to modern agriculture, farmers have doubled the production of world food supplies since 1960, tripled the output of foods like cooking oils and meats, and increased per capita food supplies in the developing world by 25 percent.

Pharmaceuticals: diagnostics, prescription drugs, vaccines, vitamins, and over-the-counter drugs for human and veterinary applications. This subsector also includes biotechnology products. Strategic investment in companies, facilities, and research and development is especially important for this subsector.

Consumer products: soaps, detergents, and cleaners, as well as toiletries and cosmetics. While consumer products are an established segment of the industry, technological innovation and product development are important due to short product life cycles.

Organization for Economic Co-operation and Development (2001), divides the chemicals industry into four groupings:

Basic chemicals (or commodity chemicals). Basic chemicals represent a mature market, which is illustrated by the fact that 46 of the top 50 highest volume chemicals in 1977 were still in the top 50 in 1993. The industry is characterized by large plants, mainly using continuous – as compared to batch – operations, with high energy consumption, low profit margins, and a high degree of cyclicality over the business cycle due to fluctuations in capacity utilization and feedstocks (or raw materials). Markets for basic chemicals are primarily in other basic chemicals, specialty chemicals, and other chemical products, as well as in other manufactured goods (textiles, automobiles, appliances, furniture, etc.) or in the processing applications (pulp and paper, oil refining, aluminum processing, etc.) (Swift, 1999).

Specialty chemicals. These chemical substances (e.g. adhesives and sealants, catalysts, coatings, electronic chemicals, plastic additives), which are derived from basic chemicals, are more technologically advanced products than basic chemicals. They are manufactured in lower volumes than basic chemicals, give higher profit margins and have less cyclicality in their business cycle. Specialty chemical products have a higher value-added because they cannot easily be duplicated by other producers or are shielded from competition by patents. Although dedicated and continuous operations are typical, there are also a growing number of plants that are general-purpose synthesis operations.

Life science products. This group includes pharmaceuticals, products for crop protection and products of modern biotechnology. Plants, generally, use batch-oriented synthesis or formulating operations where quality control and a clean environment are critical. Technological advantages are extremely important and research and development spending for this sector is the highest among all industries.

Consumer care products: This includes soap, detergents, bleaches, laundry aids, hair care products, skin care products, fragrances, etc. Consumer care products is one of the oldest segments of the chemistry business. These products are formulated products, employing what is often simple chemistry but featuring a high degree of differentiation along branding lines. Research and development expenses are rising and many of these products are becoming high-tech in nature. Consumer care products are generally formulated in batch-type operations, although some products (e.g. detergents) are manufactured in large dedicated plants. Formulating involves mixing, dispersing, and filling equipment rather than reactors for chemical conversions. We should keep in mind that the lines between these sectors can become somewhat blurred. The starting point for the chemicals industry is raw materials such as oil, coal, natural gas, air, water and minerals (see Figure 69).

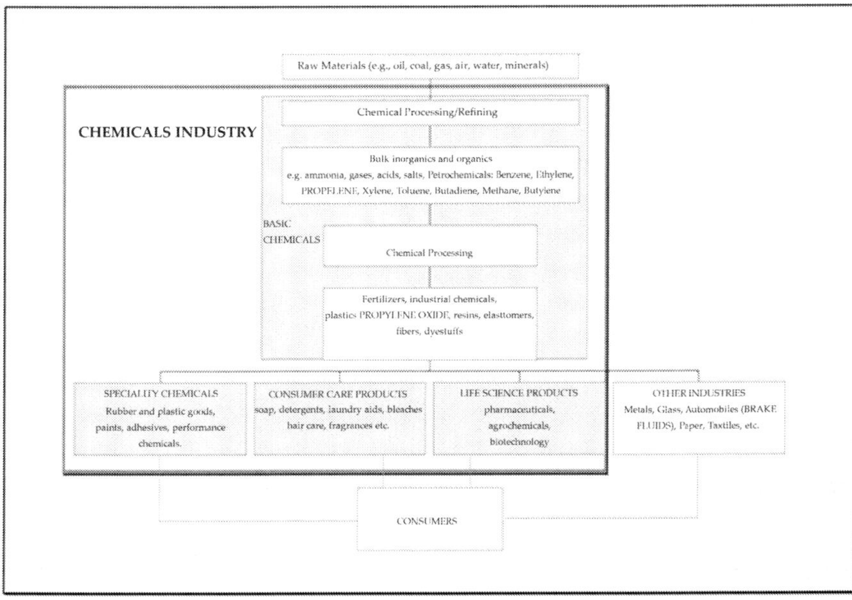

Figure 69: Overview of the Chemical Industry and Related Industries.

Source: EEA, (1995); Swift, (1999); OECD, (2001). Available at: https://www.oecd.org/env/ehs/2375538.pdf

As we can understand from the figure 69, like a tree, which starts as one trunk and then expands to multiple branches, the chemicals industry uses few raw materials to produce tens of thousands of different chemical substances. The raw materials are refined to make primary or first-level chemicals (e.g. benzene, ethylene, propylene) which are then further processed to produce second-level chemicals (e.g. dodecylbenzene, ethylene oxide, propylene oxide), and so on. These products are most often end products only as regards the chemicals industry itself; a chief characteristic of the industry is that its products nearly always require further processing before reaching the ultimate consumer. Thus, paradoxically, the chemicals industry is its own best customer. An average chemical product is passed from factory to factory (or to various units in the same factory) several times before it emerges from the chemicals industry to other industries and consumers (OECD, 2001).

Regardless of size, each plant operation follows a similar production chain. First, to start the process, raw materials or feedstocks (e.g.

propylene oxide) are brought to the plant. They can be shipped to the plant via tanker/barge, rail, pipeline, or truck, or they can be produced by the same company but at another part of the plant (OECD, 2001). After the actual processing, the chemical product is packaged and shipped to consumers (e.g. as brake fluid, antifreeze, cosmetics, tapes), or to another part of the plant, chemical company, or to another industry (e.g. automobiles, textiles, paper).

The chemicals industry is very diverse in terms of production and products, and the types of impacts it can have on the environment and human health. That is why chemical industry is one of the most regulated of all industries, it is subject to a number of requirements aimed at minimizing releases of chemical substances during manufacturing and processing.

As a major user of raw materials, both for energy consumption (7% of world energy use in 1998) (IEA, 2000) and as feedstocks, the chemicals industry can significantly impact the supply of non-renewable resources. Combustion of hydrocarbon fuel sources during manufacturing processes can lead to emissions of carbon dioxide (CO_2) – a greenhouse gas (GHG) – and volatile organic compounds (VOCs) and nitrogen oxides (NOx) which contribute to the smog formation.

The processing of the raw materials and feedstocks can result in the release of hazardous pollutants to the environment (e.g. propylene) from emission stacks, discharge pipes, flanges, waste ponds, storage tanks and other equipment. Of all the sectors of the chemicals industry, the basic chemical sector is generally the largest emitter (by volume) of such pollutants because these bulk chemicals are usually produced in high volumes at large plants. During normal operations, workers can be exposed to pollutants in a gaseous or liquid form, for example by inhaling a pollutant emitted from leaks in equipment or splashing the substance on the skin or in the eyes. Larger accidents involving chemicals can also occur due to equipment failures. Major spills can result in inadvertent releases to workers, the surrounding neighborhood or perhaps even communities and the environment at some distance from the plant (OECD, 2001).

At the end of each stage of the process, the product is transported, via pipeline, rail, barge/tanker or truck, to the next user for further processing. During transportation, accidents can occur leading to exposures of the transport workers and, perhaps, the surrounding community and environment.

The final user of a chemical product can be another chemical company, other industries, or consumers. Depending on the product, and how it is used, there also can be exposures during this end-use phase. For example, as OECD (2001) reports, chemicals such as plasticizers and stabilizers found in plastics could leak out during consumer use. Leakage of brake fluids from tanks and disposal of these substances (generally classified as hazardous waste) can cause hazardous effect on the environment.

The use of some consumer products can have a global impact, as is the case with refrigerants containing CFCs that have led to a depletion of the ozone layer. Certain chemicals released by the chemicals industry in its production processes can lead to a direct or indirect impact on the environment or human health. Exposure to certain hazardous substances – such as PCBs, DDT, PBBs, heavy metals, endocrine disrupting substances – can lead to a direct toxicological effect on man or the environment from short- or long-term exposure. Other substances – such as VOCs, NOx, SOx – are cause for concern after they react with other substances. For instance, VOCs and NOx promote the formation of smog and SOx is responsible for the formation of acid rain. CFCs are non-flammable and non-toxic, but they react with other substances in the stratosphere to destroy some of the ozone layer and this, in turn, can lead to a greater incidence of skin cancer (OECD, 2001).

Strict monitoring of pollutant emissions under the chemical industry's Responsible Care Global Charter (2016) have ensured significant reductions in the environmental impact of chemical production. However, there has been little focus on the indirect environmental effects of energy use at chemical plants. Some energy is required in the manufacturing process itself, but some may even be used as part of efforts to control pollutant emissions. For example, waste gas incineration, recycling and wastewater treatment may all lead to higher energy consumption (Wernet et.al., 2010).

The study of G. Wernet and his colleagues (2010) assessed the relative importance of energy use to the total environmental impact in the production of 99 different chemicals in Western Europe. The analysis included common inorganic chemicals, such as ammonia, chlorine and sulfuric acid, and organic chemicals, fluctuating from simple compounds, such as hexane and methanol, to more advanced dyes and pigments (Wernet et.al., 2010).

The key concept used in study was the Life Cycle Analysis. Life-cycle inventory (LCI) data were collected for each chemical using the eco-invent database and information from enterprises in the chemical industry. The inventory data included all energy and material inputs and all releases to air, land and water that occurred before and during production. All data was categorized as either 'energy-related' or associated with 'remaining causes'. Energy-related factors were further divided as 'related to transport' and 'related to heat or electricity production'. Remaining causes included direct emissions during production and waste treatment, and the provision of raw materials (Wernet et.al., 2010).

Using Life Cycle Impact Assessment (LCIA) methodology, the researchers calculated the environmental impact of manufacturing process of each chemical, including both energy-related and non-energy-related factors. Then they compared obtained results with the result when energy-related factors were removed from the analysis, so that their environmental impact could not contribute to the overall impact score (Wernet et.al., 2010).

The scientists used several types of LCIA to evaluate different aspects of environmental impact:

- Cumulative Energy Demand (CED), which evaluated the primary resources use;
- Global Warming Potential (GWP), which assessed the release of greenhouse gases;
- three Ecoindicator 99 indices that assess impacts on human health, resources, and ecosystem quality (using the hierarchic weighting scheme);
- several Ecological Scarcity indices, with measuring of ecological performance across a range of categories, including freshwater input and land occupation (Wernet et.al., 2010).

This combination of methods provides a range of viewpoints, allowing to form the optimal conclusions. As a result, energy-related factors accounted for 40-80 % of the environmental impact for most LCIA methods. The impact was particularly high for the Global Warming

Potential, for which energy-related processes were responsible for up to 80 %. On average, half of the energy-related impact could be attributed to the use of steam and other forms of heat, with the use of electricity responsible for a further 40-50 %. The remaining 5 % was attributable to transport. The study revealed that direct atmospheric pollution was responsible for less than 10 % of the total impact of the chemical industry on human health (Wernet et.al., 2010).

For these reasons, G. Wernet and his team (2010) suggest that implementing more efficient and sustainable sources of energy, such as natural gas and renewable energy, will be the most effective available tool to help reduce and monitor the environmental impact of the chemical industry in Europe.

Although the impacts from hazardous chemicals produced by the chemicals industry are complex, and sometimes open to debate, some negative effects have been well documented. Maybe, the best known example is the dichlorodiphenyl trichloroethane (DDT), an insecticide, that was developed to control a number of insect pests and is widely used in tropical countries for disease vector control (malaria, yellow fever). But it has been demonstrated that it can cause reproductive failure in eagles and other birds due to the thinning of eggshells (EEA, 1998). Certain polychlorinated biphenyls (PCBs) are relatively fire resistant and are employed primarily as cooling and insulating fluids in industrial transformers and capacitors. However, PCBs are persistent in the environment and can lead to reproductive effects in some mammalian species. The vinyl chloride monomer in polyvinyl chloride (PVC) - that is used for a range of products such as pipes, films, bottles, floors and walls - has been shown to cause cancer (EEA, 1998). Phosphates in washing powders and detergents cause eutrophication of aquatic ecosystems (EEA, 1998). Endocrine disrupting chemicals are also a concern as they interfere with the normal function of the hormonal systems of humans and animals. These properties have been found in several classes of chemicals released into the environment, such as some insecticides and fungicides, phthalate plasticizers, dioxins, and antifouling paints (OECD, 2000).

Most recently, concern has been expressed about chemicals which are characterized by persistence in the environment, resistance to degradation, and acute and chronic toxicity. Further, some of these can be transported over long distances through the atmosphere or aquatic systems and pollute areas where they have never even been used. Table 7 shows us some examples of ecological impacts of chemical industry gathered by European Environment Agency. The association/causation is assessed on the scale: 1 = no observed association, 2 = suspected association, 3 = weak association, 4 = clear association, 5 = significant association.

Table 7: Examples of ecological impacts and possible causes

Observation/impact	**Sensitive species**	**Substance**	**Association/ causation**
Large scale			
Eggshell thinning	guillemot, eagle, osprey, peregrine falcon	DDT	5
Reproduction	seal, otter	PCB	4
Skeleton malformation	grey seal	DDT, PCB	2
Pathological changes	seal	PCB, DDT, metabolites	3
Reproduction	mink	PCB	5
Reproduction	osprey	DDT, PCB	4-5
Reproduction	eagle	DDT, PCB	2-3
Reproduction	salmon	chlorinated substances	2
Large scale - pulp and paper industry			
Induction of metabolising enzymes	perch	chlorinated/un-chlorinated organic mixture/ Dioxin compound	3
Local/regional - pulp and paper industry			
Induction of metabolising enzymes	perch	chlorinated/un-chlorinated organic mixture/ Dioxin compounds	3-4
Spine malformations	four-horned sculpin	chlorinated/un-chlorinated organic mixture	3-4
Local, forest industry			

Induction of metabolising enzymes	perch	chlorinated/un-chlorinated organic mixture Dioxin compounds	4-5
Spine malformations	four-horned fish	chlorinated/un-chlorinated organic mixture	4-5
Larvae damages	sea mussel	chlorinated/un-chlorinated organic mixture	3

Source: European Environment Agency (EEA), (1998). Ecological and Human impacts. In: Chemicals in the European Environment: Low Doses, High Stakes? Available at: https://www.eea.europa.eu/publications/NYM2/page008.html

In addition, there has been a growing focus in countries on investigating the impacts of chemicals on children's health (e.g. certain chemicals like lead, mercury and polychlorinated biphenyls which may have harmful and possibly permanent neurological effects on children). A child's nervous system, reproductive organs, and immune system grow and develop rapidly during the first months and years of life. As organ structures develop, vital connections between cells are established. These delicate developmental processes in children may easily and irreversibly be disrupted by toxic environmental substances, such as lead (OECD, 2000).

During its development, chemical industry has become more complex and sophisticated in its design and use of technology to reduce environmental impact. With evolution of control technologies, the cost of reducing the last remaining amounts of pollutants emitted grew at a much faster rate than the amount of emissions reduced. As a result, companies began to examine their entire production process to find more cost-effective ways to minimize the amount of substance that reaches the end of the pipe. At the same time, governments began – and continue – to focus more on policies that prevent rather than control pollution.

One of the most promising new approaches used by the chemicals industry is to encourage 'sustainable chemistry' (OECD, 2001). Within the broad framework of sustainable development, efforts are being made to maximize resource efficiency through energy and non-renewable resource conservation, risk minimization, pollution prevention, waste reduction at all stages of a product's lifecycle, and the development of products that are durable and can be reused and recycled. Sustainable chemistry strives to accomplish all this through the design, manufacture and use of efficient and effective, and more environmentally benign, chemical products and processes.

With the increasing reliance on computers, and the greater availability of laboratory test data on some chemicals, more researchers are developing models that can predict the effects other chemicals may have if man or the environment were exposed. Such models can reduce the need for animal testing, and generate more data faster than is possible today. As many countries have collected considerable test data on new and existing substances, governments may wish to use these data more widely to compare results predicted by models with actual results found in laboratories, and possibly co-ordinate the results of this work to develop new models (US EPA, 2017).

Organization for Economic Co-Operation and Development Considering pointed out three main groups of issues that deserve more attention in the chemical industry development:

Creating a holistic chemical safety approach: Developing such an approach means:

- improving the knowledge base for the design of safe chemicals (building on current sustainable chemistry efforts);
- finding ways to better evaluate and manage the risks resulting from the release of chemicals from products; this will involve further developing lifecycle assessment techniques and paying more attention to the risks for specific, sensitive groups of the population (such as children) or for sensitive ecosystems;
- finding the right means to balance the efficacy of products with their overall "environmental and health performance" (including resource use, safety during production and releases from product use) which will require an Integrated Product Policy;
- effective implementation of an Extended Producer Responsibility regime to ensure that industry takes the Responsible Care concept a step further; this would also mean that large multinational companies which apply Responsible

Care will make efforts to assist small- and medium sized enterprises in implementing more comprehensive pollution prevention, resource efficiency and safety policies;

- further developing and extending the concept of the chemicals industry as a "service industry"; as these chemical "service" companies focus on providing a function for their customers rather than a product, they have more opportunities to reduce risks at the production, use and waste disposal stage; chemicals users and governments could work together to develop, apply, and propagate this concept.

Managing the safety aspects of globalization in the chemicals industry This will involve:

- supporting the development of the chemical safety regimes in non-OECD countries. One way of doing this would be to involve them more closely in OECD work;
- finding ways to respond to potential negative impacts on the environment due to the fast pace of restructuring and reorganization in multinational enterprises. If national structures cannot adequately deal with this, closer international co-operation might be needed to develop efficient and fast international information exchange and control systems.

Environmental democracy in chemical safety: Governments and industry must work closely together to ensure chemical safety, and trade unions, the public and workers must play an active role in the decision-making process so that there will be greater transparency, and better and fairer policies can be implemented. This will call for:

- information on emissions and the safety of chemicals needs to be made widely available; public right-to-know should be strengthened as much as possible and yet be compatible with justified business interests;
- information should be made available in a way that will make it possible for public interest groups and the public to understand the implications of the information in terms of risks to human health and the environment;
- efforts should be made to educate the public about chemical

safety and, where necessary, to provide public interest groups with resources so that they can play an equitable role in policy discussions related to chemical safety (OECD, 2001).

TRANSPORT

Transport (transportation) is the movement of people and goods from one location to another. Transport enables trade between people, which is essential for the development of civilizations. Transportation as industrial sector can be divided into infrastructure, vehicles and operations. Modes of transport include air, rail, road, water, cable, pipeline, and space.

Transport infrastructure consists of the fixed installations: roads, railways, airways, waterways, canals, and pipelines; and terminals: airports, railway stations, bus stations, warehouses, trucking terminals, refueling depots (including fueling docks and fuel stations) and seaports. Terminals may be used both for interchange of passengers and cargo and for maintenance.

Talking about transport from the Industrial Ecology point of view, we mean vehicles. A vehicle is a non-living device that is used to move people and goods. The vehicle moves along with the cargo and riders (drivers and passengers), providing its own propulsion. This is most commonly done through a steam engine, combustion engine, electric motor, a jet engine, or a rocket, though other means of propulsion also exist. Vehicles also need a system of converting the energy into movement; this is most commonly done through wheels, propellers, and pressure. Vehicles traveling on the transportation networks include automobiles, bicycles, buses, trains, trucks, people, helicopters, watercrafts, aircrafts, and space crafts.

Transportation sector of industry causes diverse environmental impacts on the all components of environment and human health. Let us analyze main environmental issues of transportation process.

Air pollution: As transport uses 75 % of extracted fossil fuels in form of petroleum, it is responsible for the greatest amounts of many air pollutants, especially greenhouse gases. More than 80 % of total carbon dioxide and 55 % of methane emissions are caused by transport. Same results are derived for the case of sulfur and nitrogen oxides with a share of 65 % and 61 % of total emissions respectively, occurring during the operation stage. Around 86 % of the total number of hydrocarbons is emitted also during the utilization stage of a gasoline vehicle (Poulikidou, S., 2010).

Direct exhaust emissions of hazardous pollutants like carbon monoxide, nitrogen oxides and particulates are important contributors to many environmental problems such as smog creation and biodiversity disturbances (Lane, 2006). Those emissions affect the natural ecosystem at a more local level. At a global level, the transportation sector causes the greenhouse gas emissions and global climate change since the utilization of the vehicle and other activities related to the sector are responsible for a significant amount of carbon dioxide emitted to the atmosphere (Poulikidou, S., 2010). More than 16% of human made CO_2 emissions are caused by road transportation in general whereas another 7% to 10% is assigned to passenger cars only (ACEA, 2016).

US Environmental Protection Agency informs that people who live, work or attend school near major roads appear to have an increased incidence and severity of health problems associated with air pollution exposures related to roadway traffic. Children, older adults, people with preexisting cardiopulmonary disease, and people of low socioeconomic status are among those at higher risk for health impacts from air pollution near roadways. The health effects include:

- Higher rates of asthma onset and aggravation;
- Cardiovascular disease;
- Impaired lung development in children;
- Pre-term and low-birthweight infants;
- Childhood leukemia; and
- Premature death (EPA, 2016).

Land consumption and landscape damage: The provision of land-based transport requires the direct utilization of land. Previous land uses, such as forestry, agriculture, housing, and nature reserves, may be displaced, and zones adjacent to the new development rendered unsuitable for wide range of activities (Raghav, 2017).

The latter aspect is true of pipelines carrying volatile materials (such as natural gas), for example, where a corridor of land along the route must be kept undeveloped for safety reason, even if the pipeline itself causes no direct consumption of land. Existing severance may seriously restrict the movement of people and animals between previously contiguous areas, with consequences for the quality of community life and the functional integrity of ecosystems (Klymenko, 2000).

Airports demand big areas of land to serve all their facilities, so they create hazardous effects in their particular location. Some environmental effects, especially from non-motorway roads, are only partial. Airports cause increasing of traffic density and speed, and that increases the danger of pedestrian crossings on the same level. The use of light-controlled crossings, road tunnels or viaducts can reduce severance, though the latter introduce significant visual impact, and both solutions are costly.

Land consumption is not just a direct effect of transport development; it may also occur indirectly as land is utilized for the extraction of the raw materials (principally aggregate) required for construction. An average of 76,000 tons of road mixture (aggregate) is required per kilometer of road lane, and approximately 90 million tons of aggregates are used in the UK every year in the construction and repair of roads (Royal Commission on Environmental Pollution, 1997).

The important environmental impact of transport-related land loss and land use is the visual pollution (decline in visual amenity or aesthetic attraction of the landscape). Visual impact may be essentially linear in nature for road, rail, and inland waterway developments, or nodal in character for the large terminal installations, such as sea ports and airports.

Water pollution: Motor vehicles and transport infrastructure are a source of water pollution and hydrologic disruptions. Water pollution effects include (EPA, 1999):

- Crankcase oil drips and disposal.
- Road de-icing (salt) damage.
- Roadside herbicides.
- Leaking underground storage tanks.
- Air pollution settlement.

Hydrologic impacts are characterized by (Forman, et al, 2003):

- Increased impervious surfaces.
- Concentrated runoff, increased flooding.
- Loss of wetlands.
- Shoreline modifications.
- Construction activities along shorelines.

These impacts impose various environmental costs such as polluted surface and ground water, contaminated drinking water, increased flooding and flood control costs, wildlife habitat damage, reduced fish stocks, loss of unique natural features, and aesthetic losses (Victoria Transport Policy Institute, 2015).

Helen Pressley (1991) described that an estimated 46% of US vehicles leak hazardous fluids, including crankcase oil, transmission, hydraulic, and brake fluid, and antifreeze, as indicated by oil spots on roads and parking lots, and rainbow sheens of oil in puddles and roadside drainage ditches. An estimated 30-40% of the 1.4 billion gallons of lubricating oils used in automobiles are either burned in the engine or lost in drips and leaks, and another 180 million gallons are disposed of improperly onto the ground or into sewers (Pressley, 1991). Runoff from roads and parking lots has a high concentration of toxic metals, suspended solids, and hydrocarbons, which originate largely from automobiles (Bannerman et.al., 1993).

Road de-icing salts cause significant environmental and material damages, causing soil salinization and degradation, loss of harvest, vehicles corrosion. Hydrologic impacts of roads and parking facilities are serious as they concentrate storm water, increasing flooding, scouring and siltation effects, reducing surface and groundwater recharge which lowers dry season flows, and create physical barriers to fish.

Ecosystem degradation: The degradation of terrestrial and aquatic ecosystems, as measured by indicators such as reduced habitat/species diversity, primary productivity, or the areal extent of ecologically valuable plant and animal communities, provides one of the most emotive aspects of the tension between transport development and environmental quality (Raghav, 2017).

Road network physically divides natural and semi-natural ecosystems and, in that way, destroy ecologic corridors of migration, inhibiting the movement of animal and plant species across transport lines. Associated reduction in size can threaten the biodiversity and viability of the smaller ecosystems. The direct consequence of road transport is the death of individual animals through collision with vehicles. However, the indirect or secondary effects of transport development may also be responsible for many adverse impacts on wildlife, including those associated with air, water, and noise pollution.

Noise pollution: The European Commission acknowledges that traffic noise is one of the main local environmental problems in Europe and the source of an increasing number of complaints from the public. The negative effects of transport noise have been known for years, but considering pressing of air pollution problems, noise has often taken a lack of attention. The TERM 2008 report of European Environment Agency "Transports at a crossroads" painted a bleak picture – 55% of those living in urban areas with more than 250 000 inhabitants in the European Union 27 member states (67 million people) endure daily road noise levels above the lower EU benchmark for excess exposure (Health and Environment Alliance, 2011).

As Health and Environment Alliance (2011) admits, the growing body of research reveals, transport noise can cause sleep disturbance, cardiovascular disease, elevated hormone levels, psychological problems and even premature death; studies on children have identified cognitive impairment, worsened behaviour and diminished quality of life.

Waste generation: According to data published by the European Commission, in terms of material losses and waste accumulation, 8 to 9 million tons of waste are generated from end-of-life vehicles every year. The environmental impact caused by the treatment of old vehicles and vehicle components has been reduced significantly the latest years especially after the implementation of several regulation measures. However, there is still a flow of materials, sometimes scarce or hazardous, that end up in landfills. In addition to that, end-of-life treatment efficiency is negatively affected by the increasing amounts of newly developed materials that have limited recyclability (i.e., plastics and carbon composites) (Poulikidou, S., 2010).

Social issues: Transport, giving advantages of mobility to people, is also the serious reason of social-ecological issues. When it comes to commuting to work in U.S. cities, people living in New York have to travel for the longest period of time (StatisticaCharts, 2014). In fact, almost one third of New Yorkers have a commute lasting 60 minutes (Figure 70).

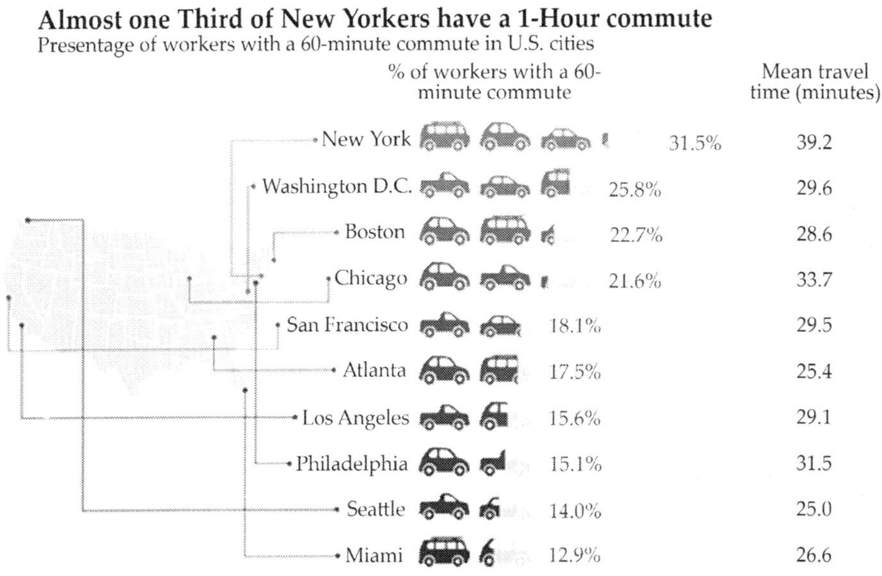

Figure 70: Percentage of Workers with a 60-Minute Commute in U.S. Cities.

Source: Bloomberg, StatisticaCharts, (2014). Available at: https://www.statista.com/chart/1754/daily-commuting-time-in-us-cities/

As it is shown at the Figure 70, the mean travel time in minutes for New Yorkers is 39.2. Washington D.C. has the second longest commute in the United States. 25.8 % of commuters in the American capital have a one hour journey to work. This number is 22.7 and 21.6 % in Boston and Chicago respectively. When it comes to commuting to work by car, Washington D.C. has the worst delays in the United States (StatisticaCharts, 2014). Therefore, transport steel so precious time from people life.

Another aspect of social impact is the real cost of transporting which calculation includes not only the money people pay from their wallets actually for transporting, but also cost of physical fitness, cost to the environment, and community cost. Discourse Media presented analysis of true cost of commuting (Figure 71).

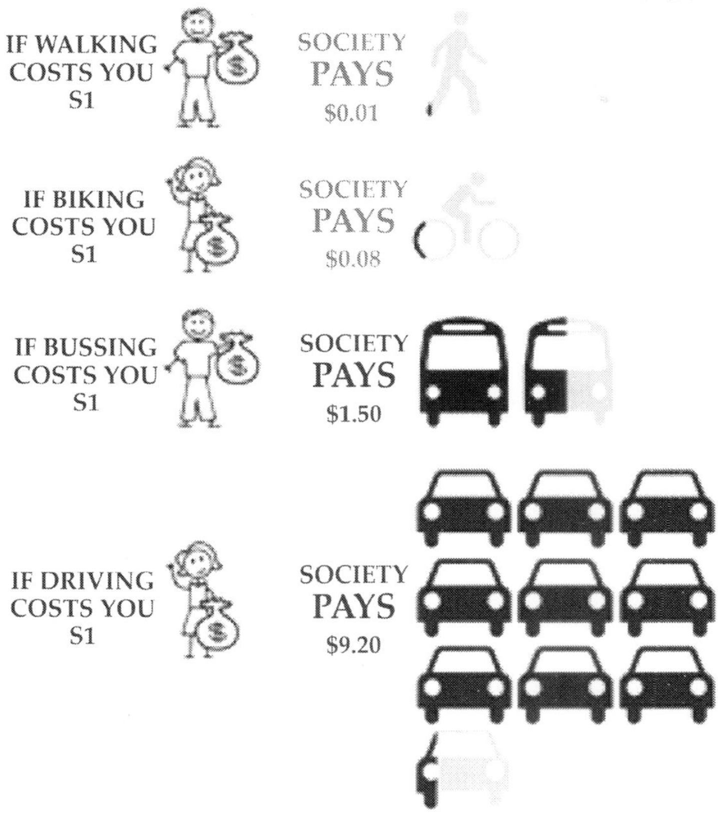

Figure 71: True Cost to Commute.

Source: Discource Media, data by George Poulos (2017); LEED Bicyclology. (2017). What is your TRUE Cost to Commute? LEED Bicyclology by LEED Bicycle Solutions, LLC. Available at: https://blog.e-bikerig.com/2015/07/28/what-is-your-true-commute-cost/

Commuting by car costs society 35 cents per mile because of all the public's costs as a whole for the infrastructure etc. On the contrary, commuting by bike saves society 9 cents per mile. Most Americans live within 8 miles from work, which is deemed the 'commutable' distance by bike. This means that if those of us that live within 8 miles of work would commute by bicycles, we could save up to $11.52 individually and could save society $7.04 for a grand total savings of $18.56 every day (LEED Bicyclology, 2017).

Considering powerful environmental impacts of transport, changes in transportation policy and infrastructure should be done for reductions in health risks related to air and noise pollution exposures and traffic injuries. World Health Organization (2011) admits that cycling, walking and using rapid transit/public transport can greatly enhance levels of physical activity, helping prevent a range of chronic diseases including cancer, heart disease and diabetes. In addition to these health implications, the transport sector offers major potential for climate change mitigation. Mitigation policies therefore need to be examined in light of both these impacts to identify mitigation strategies with the greatest potential to reduce climate change and gain health co-benefits (WHO, 2011).

US Environmental Protection Agency (2016) developed the system of technologies to reduce transportation air pollution (Figure 72).

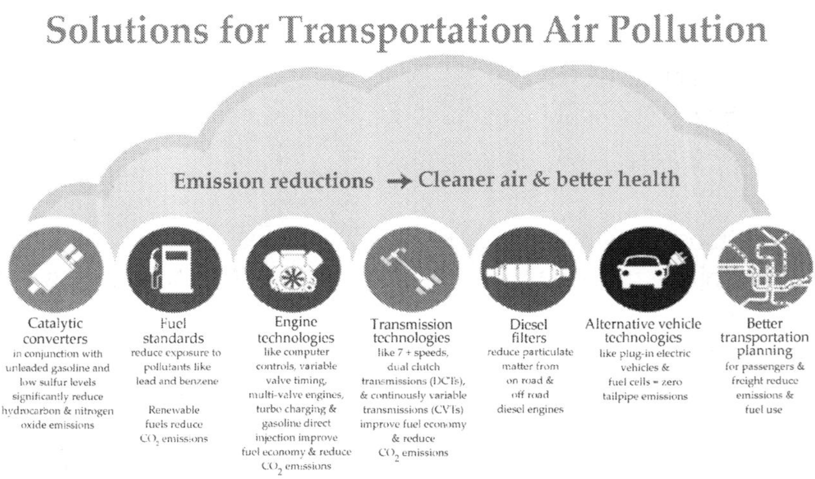

Figure 72: Technological Solution for Transportation Air Pollution.

Source: EPA, (2016). Available at: https://www.epa.gov/air-pollution-transportation/learn-about-air-pollution-transportation

World Health Organization, (2011) undertook a review of potential health co-benefits of transport mitigation strategies:

- A shift to active transport (walking and cycling) and rapid transit/public transport combined with improved land use can yield much greater immediate health 'co-benefits' compared with improving fuel and vehicle efficiency.
- Potential health gains of a shift from private motorized transport to walking, cycling and rapid transit/public transport include reduced respiratory and cardiovascular disease from air pollution and less exposure to traffic injury risks and noise stress. In addition, large benefits are expected from increased physical activity leading to the prevention of obesity, diabetes, heart disease and cancer, as well as greater health equity achieved by better access to goods and services among groups without private motor vehicles.
- Land use planning that increases density and diversity of uses must be a core principle in design of mitigation strategies with greater health co-benefits. Emphasis on 'proximity planning' can improve access to goods and services via walking, cycling and public transport infrastructure. Reducing the need to travel by car will reduce transport-related emissions and lead to long-term co-benefits to health and climate change.
- Investments in active transport and rapid transit/public transport can assist the transport sector to achieve its own objectives by reducing congestion and the need to fund costly road infrastructure. Transport systems with strong walking, cycling and public transport provision are also less vulnerable to future interruptions in supply of oil or other fuels (WHO, 2011).

Conserve Energy Future (2017) proposes the strategy of Green Transportation, which means any kind of transportation practice or vehicle that is eco-friendly and does not have any negative impact on the immediate environment. Green transportation requires dedicated public awareness and participation, management of privately owned vehicles and manufacturing of vehicles that utilize renewable sources of energy such as wind, solar, biofuels and hydroelectricity. Modes of green transportation include (Conserve Energy Future, 2017):

Bicycle: Using bicycle to commute is key mode of green transportation. Shifting from driving a car to riding a bicycle enormously contributes to ridding the atmosphere of greenhouse gasses. Walking is also considered a great green transportation mode, but bicycle has far reaching benefits because it's faster (Conserve Energy Future, 2017).

Electric vehicles: Some notable types of electric vehicles include cars, motorcycles, lorries, trains, boats, and scooters. Electric vehicles powered entirely by electricity do not emit any air pollutants in urban areas where they operate. On the other hand, toxic emissions might be produced by powerplants generating the electricity. Still, the power can be generated using the renewable energy sources like geothermal, hydroelectric, solar power and wind turbines (Conserve Energy Future, 2017).

Hybrid cars: Design of hybrid car suppose the automatic recharging of car battery by converting energy in the course of braking. Greenhouse gases emissions in hybrid cars are tremendously low; amount of other air pollutants can range from 26% – 90% lower compared to standard cars. Experts inform that hybrid cars cut down health-threatening emissions by over 90%. While hybrid cars have benefits from the emission point of view, their batteries have some environmental impacts as the caustic substances lighting up the batteries must be carefully and well disposed of (Conserve Energy Future, 2017).

Multiple occupant vehicles: When some group of people habitually driving in similar direction, they can make use one vehicle. Instead of 5 individuals driving their own cars in the same direction, it's a lot more economical (saves money and fuel) and ecologically efficient to make use of a single car to take all of you to the destination. Multiple occupant vehicles, or carpools, reduce the number of vehicles on roads, thus, minimizing levels of pollution. Multiple occupant vehicle is an eco-friendly and favorable mode of green transportation (Conserve Energy Future, 2017).

Hybrid buses (Public Transportation): Some of the best innovative green buses have appeared at the streets during last years. A classic example is the Mercedes-Benz Ciatro G BlueTec Hybrid Bus, which utilizes 4 electrical wheel hub motors together with automotive lithium-ion batteries. The lithium-ion battery has the ability to store up huge amounts of energy derived from a diesel generator. The bus also generates energy during breaking (Conserve Energy Future, 2017).

Green trains: Trains are getting progressively greener with hybrid locomotives and other innovative green technologies as world governments now dedicated to supporting green transportation. The innovative hybrid locomotives exploit similar technologies applied in hybrid cars. The up-to-date electric trains use the electrified third rail, overhead lines or devices that store up energy like fuel cells and batteries. Electric trains travel at tops speeds of more than 200 mph, yet maintaining high safety levels (Conserve Energy Future, 2017).

Service and freight vehicles: These kind of transport attributes to about 9% of the total air pollutants emissions. Utilizing electricity and biofuels instead of the regular fossil fuel sources in services and freight vehicles can decrease the environmental risks of transportation (Conserve Energy Future, 2017).

6

CHAPTER

INDUSTRIAL ENVIRONMENTAL DISASTERS

In this book, we have already analyzed the background of Industrial Ecology, its main objectives and key concept, we understood main types of industrial environmental impacts, their causes and consequences, we investigated how different industrial sectors influence the environmental and human health and what new technologies we have to eliminate that hazardous effects. In the last chapter let's look ones more at the global industrial environmental disasters of the past, so we will always remember them and never let such tragedies danger the humanity again.

LONDON'S KILLER FOG

With the advent of industry, London's population was accustomed to seeing foggy, pollution laden air. In 1952 however, this pollution took a tragic turn. That winter, the weather was cold and residents burned more coal in their fireplaces to alleviate the chill (Disasterium, 2015).

Christopher Klein (2012) describes that for five days in December 1952, London gasped for air. A toxic fog smothered the British capital, paralyzing the city and blackening out the sun. The smoke laced with sulfur dioxide, nitrogen oxides and soot, and left London encased in a black cloud of near total darkness and killed over 12,000 people, and more than 150,000 people had been hospitalized.

Smog was nothing new in London, but this particular 'pea souper' quickly thickened into a poisonous brew unlike anything the city had ever experienced (Figure 73). A high-pressure system parked over London caused a temperature inversion—with the air 1,000 feet above the surface warmer than that at ground level—that prevented the smoke from the record amount of coal being burned into the skies from rising. And with nary a breeze to be found, there was no wind to disperse the soot-laden soup (Klein, 2012).

Figure 73: London's Killer Fog.

Source: Klein, C., (2012). The Killer Fog That Blanketed London. History in the Headlines, December 6, 2012. Available at: http://www.history.com/news/the-killer-fog-that-blanketed-london-60-years-ago

The 30-mile-wide noxious smog, teeming with acrid sulfur particles, reeked like rotten eggs. It was so dense that residents in the Isle of Dogs section of the city reported they were unable to see their feet as they walked. For five days, the Great Smog paralyzed London and crippled all transportation, except for its Underground. Boat traffic on the Thames came to a halt. Flights were grounded and trains cancelled. Even during the middle of the day, drivers turned on their headlights and hanged their heads out the windows in an attempt to inch ahead through the yellow gloom. Many found the exercise futile and abandoned their cars. Conductors grasping flashlights and torches walked in front of the iconic double-decker buses to guide drivers nosing down the city streets. Wheezing pedestrians groped their ways around the city's neighborhoods and tried not to slip on the greasy black ooze that coated sidewalks. By the time they returned home, with their faces and noses blackened by the air, Londoners resembled coal miners (Klein, 2012).

Despite its lethal nature, the exact cause and nature of the killer fog has largely remained a mystery. Recently, a team of researchers has determined the likely reasons for its formation. As Kacey Deamer (2016) informs, researchers have for a long time connected emissions from burning coal with the killer fog, but the specific chemical processes that led to the deadly mix of pollution and fog were not fully understood. To determine what turned the fog into a killer, an international team of scientists from China, the U.S. and the U.K. recreated the fog in a lab using results from laboratory experiments and atmospheric measurements from Beijing and Xi'an, two heavily polluted cities in China (Deamer, 2016).

Study lead author Renyi Zhang, an atmospheric scientist at Texas A&M University, said that sulfate was a big contributor to the deadly London fog. Sulfuric acid particles, which formed from the sulfur dioxide that was released from the burning of coal, were also a component of the fog. The question was: how did sulfur dioxide get turned into sulfuric acid? "Our results showed that this process was facilitated by nitrogen dioxide, another co-product of coal burning, and occurred initially on natural fog," Zhang said in a statement. "Another key aspect in the conversion of sulfur dioxide to sulfate is that it produces acidic particles, which subsequently inhibits this process". The natural fog contained larger particles, Zhang explained, with the smaller acidic particles evenly distributed throughout. When those fog particles evaporated, an acidic-haze was left covering the city. (Deamer, 2016).

"DOOR TO HELL"

Turkmenistan is seventy percent desert – the Karakum Desert, to be exact In the late sixties, the Soviet Union sent exploration teams across the continent to locate deposits of gas and oil. By 1971 Soviet scientists discovered a cache of oil reserves near the town of Derweze in the middle of Karakum Desert. The gas reserve found there was one of the largest in the world.

But when a drilling rig collapsed it created a crater, and large amounts of methane were released. When the oilmen attempted to burn off the methane, it started a fire in a "door to hell" that still burns over forty years later.

A camp was established, a drilling rig was quickly set up, and operations began. As the drilling started, the petrochemical scientists started to estimate the quantity of gas reserves available at the site. Initial estimates were positive, and when the Soviet drilling rigs confirmed their findings, production was increased to full capacity and they began the storage of the gas (Sometimes Interesting, 2013).

The ground beneath the drilling rig and camp collapsed into a wide crater and disappeared. No lives were lost in that incident. However, large quantities of methane gas were released, creating an environmental problem and posing a potential danger to the people of the nearby villages. Fearing the release of further poisonous gases from the cavern, the scientists decided to burn it off. They thought that it would be safer to burn it than to extract it from underground through expensive methods. Environmentally, gas firing is the next best solution when the circumstances are such that it cannot be extracted for use (Miningawareness, 2013). According to the Soviet practice, scientists' expectations were that the gas would burn within days, but it is still burning, more than four decades after it was set on fire (Figure 74).

Figure 74: Panorama of the Gas Site Burning, 2011.

Source: Tormod Sandtorv (2011). Available at: https://en.wikipedia.org/wiki/Gates_of_hell#/media/File:Darvasa_gas_crater_panorama.jpg

The name, 'Door to Hell', was given to the field by the locals, referring to the fire, boiling mud and orange flames in Derweze's large crater with a diameter of 70 metres (230 ft). The hot spots range over an area with a width of 60 meters (200 ft) and to a depth of about 20 meters (66 ft). As of 2013, the hole is still burning. It has now become an international tourist destination (Miningawareness, 2013).

THREE MILE ISLAND NUCLEAR INCIDENT

The Three Mile Island Unit 2 (TMI-2) reactor, near Middletown, Pennsylvania, USA, partially melted down on March 28, 1979. This was the most serious accident in U.S. commercial nuclear electricity generating history, although its small radioactive releases had no detectable health effects on plant workers or the public. While little radiation was released from the accident thanks to a working containment system, this accident became the rallying call for fears about the nuclear power industry (Disasterium, 2015).

As US Nuclear Regulatory Commission (US NRC), (2014) informs, the accident began about 4 a.m. on Wednesday, March 28, 1979, when the plant experienced a failure in the secondary, non-nuclear section of the plant (Figure 75). Either a mechanical or electrical failure prevented the main feedwater pumps from water entering the steam generators that usually take heat away from the reactor core. That caused the plant's turbine-generator and then the reactor core itself to automatically shut down. Immediately, the pressure in the primary system (the nuclear section of the plant) began to increase. In order to control rising pressure, the pilot-operated safety valve opened. The valve should have closed when the pressure fell to proper levels, but it remained opened. At that point of time, instruments in the control room still indicated to the plant staff that the valve was closed. As a result, the plant staff was unaware that cooling water was driving out of the stuck-open valve (US NRC, 2014).

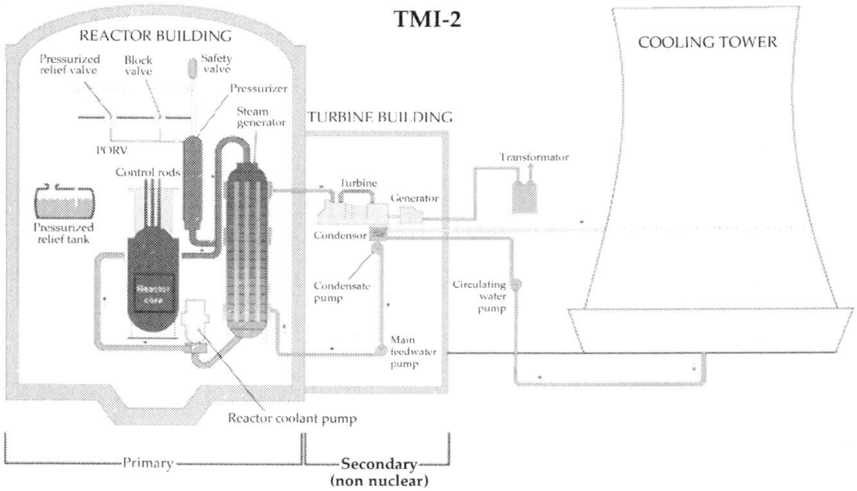

Figure 75: Three Mile Island Plant Diagram.

Source: US NRC, (2014). Available at: https://www.nrc.gov/reading-rm/doc-collections/fact-sheets/3mile-isle.html

As cooling water flowed from the nuclear section through the valve, other instruments available to reactor operators provided inadequate information. There was no tool that showed how much water covered the core. As a result, plant staff presumed that as long as the pressurizer water level was high, the core was properly covered with coolant. As alarms rang and warning lights flashed, the operators did not understand that the plant was facing a loss-of-coolant accident. A series of actions that were applied by personnel made conditions worse. The water escaping through the stuck valve decreased primary system pressure so much that the reactor cooling pumps had to be turned off to prevent hazardous vibrations. To prevent the pressurizer from filling up entirely, operators reduced emergency coolant pumping in to the primary system. These actions starved the reactor core of cooling water, causing it to overheat (US NRC, 2014).

Without the proper water flow, the nuclear core overheated to the point at which the zirconium cladding (the long metal tubes that cover uranium pellets) ruptured and the fuel pellets began to melt. Although TMI-2 experienced a severe core meltdown, TMI-2's containment building remained undestroyed and held almost all of the accident's radioactive material (US NRC, 2014). Later, it was found that about half of the core melted during the early stages of the accident (History, 2017).

US NRC (2014) admits that the small releases of radioactive gases were measured off-site by the late morning of March 28 local authorities immediately took steps to try to gain control of the reactor and ensure adequate cooling to the core. The NRC's regional office promptly forwarded the first team of inspectors to the site; other governmental agencies, such as the Department of Energy and the Environmental Protection Agency, also mobilized their response groups (US NRC, 2014).

By the evening of March 28, the reactor core seemed to be adequately cooled and the reactor appeared to be stable. But new problems occurred by the morning of Friday, March 30. A significant release of radiation from the plant's auxiliary structure, performed to relieve pressure on the primary system and avoid curtailing the flow of cooling water to the core, caused a great deal of confusion and consternation. During a short time period, chemical reactions in the melting uranium fuel created a large hydrogen bubble in the dome of the pressure vessel, the container that holds the reactor core. NRC officials were concerned that the hydrogen bubble might burn or even explode and rupture the pressure vessel. In that case, the core would fall into the containment building and perhaps cause a break of containment. On Sunday, April 1, experts determined that the bubble could not burn or explode because of the absence of oxygen in the pressure vessel, so the crisis ended. In addition, by that time, the utility team had greatly reduced the size of the bubble (US NRC, 2014).

In an atmosphere of growing concerns about the condition of the plant, the governor of Pennsylvania, Richard L. Thornburgh, announced the medical advice for pregnant women and pre-school-age children within a five-mile radius of the plant to leave the area. Massive demonstrations followed the accident, culminating in a rally in New York City that attracted upward of 200,000 people. By the mid-1980s, the construction of nuclear power plants in the United States had virtually ceased (History, 2017).

According to the US NRC (2014), approximately 2 million people around TMI-2 during the accident are estimated to have received an average radiation dose of only about 1 millirem above the usual background dose. To better understanding, we should admit that exposure from a chest X-ray is about 6 millirem and the area's natural radioactive background dose is about 100-125 millirem per year for the area. The maximum personal dose caused by accident at the site boundary was less than 100 millirem above background. In the months following the accident, critical issues were raised about possible adverse effects from radiation on human, animal, and plant life in the nuclear plant area. Government agencies monitoring the area collected the database of environmental samples of air, water, milk, vegetation, soil, and food. It was reported that very low levels of radionuclides could be attributed to releases from the accident (US NRC, 2014).

Analyzing possible health effects, Nuclear Energy Institute, (2014) informed that the TMI 2 accident caused no injuries, and at least a dozen epidemiological studies conducted since 1981 have found no discernible direct health effects to the population in the vicinity of the plant. Decades of research and scientific studies have shown no negative health effects on the population surrounding the plant.

People who suffered financial losses as a result of the precautionary evacuation following the incident were promptly paid, demonstrating the effectiveness of the industry's liability insurance protection under the Price-Anderson Act. In addition, businesses were compensated for loss of revenue, and the state and local community were compensated for the expenses incurred during the response to the threat (Nuclear Energy Institute, 2014). An investigation by the Nuclear Regulatory Commission claimed that the amount of radioactivity released was not a health threat, but antinuclear activists and many local citizens disputed this (History, 2017).

The cleanup of the damaged nuclear reactor system at TMI-2 took nearly 12 years and cost approximately US$973 million. The cleanup was uniquely challenging technically and radiologically. Approximately 100 tons of damaged uranium fuel had to be removed from the reactor vessel – all without hazard to cleanup workers or the public. Defueling the TMI-2 reactor vessel was the heart of the cleanup. The damaged fuel remained underwater throughout the defueling. In October 1985, after nearly six years of preparations, workers standing on a platform atop the reactor and manipulating long-handled tools began lifting the fuel into canisters that hung beneath the platform. In all, 342 fuel canisters were shipped safely for long-term storage at the Idaho National Laboratory, a program that was completed in April 1990. It was put into dry storage in concrete containers.TMI-2 cleanup operations produced over 10.6 megaliters of accident-generated water that was processed, stored, and ultimately evaporated safely (World Nuclear Association, 2012).

World Nuclear Association (2012) admits that TMI-2 today is in long-term monitored storage. No further use of the nuclear part of the plant is anticipated. Ventilation and rainwater systems are monitored. Equipment necessary to keep the plant in safe long-term storage is maintained.

Three Mile Island accident has changed both the nuclear industry and the nuclear regulatory rules. Public fear and distrust increased, Nuclear Regulatory Commission's regulations and oversight became broader and more robust, and management of the plants was scrutinized more carefully. Deep analysis of the Three Mile Island accident's events identified problems and led to permanent and sweeping changes in how NRC regulates its licensees – which, in turn, has reduced the risk to natural ecosystems and public health. Major changes that have occurred since the accident are listed below (US NRC, 2014):

- Upgrading and strengthening of plant design and equipment requirements such as fire protection, piping systems, support feedwater systems, containment building isolation, reliability of individual components (pressure relief valves and electrical circuit breakers), and the ability of nuclear plants to shut down automatically;
- Identifying the critical role of human performance in plant safety led to revamping operator training and staffing requirements, followed by improved instrumentation and controls for operating the plant, and establishment of fitness-for-duty programs for plant workers to guard against alcohol or drug abuse;
- Enhancing emergency preparation, including requirements for plants to immediately notify NRC of significant events and an NRC Operations Center staffed 24 hours a day. Licensees now test drills and response plans several times a year, and state and local agencies participate in such trainings with the Federal Emergency Management Agency and NRC;

- Integrating NRC observations, monitoring, and conclusions about licensee performance and management efficiency into a periodic, public report;
- Having senior NRC managers regularly analyze plant operating for nuclear plants that need significant additional regulatory attention;
- Intensifying NRC's resident inspector program – first authorized in 1977 – to have at least two inspectors live nearby and work exclusively at each plant in the U.S. to provide daily observation of licensee adherence to NRC regulations;
- Expanding performance-oriented and safety-oriented inspections, and the use of risk assessment to identify weak points for every plant to severe accidents;
- Strengthening and restructuring enforcement staff in a separate office within the NRC;
- Establishing the Institute of Nuclear Power Operations, the police making group for nuclear industry, and formation the Nuclear Energy Institute to provide a unified industry approach to generic nuclear regulatory issues, and interaction with NRC and other government agencies;
- Mitigation of accident conditions, radiation levels monitoring and plant status evaluation through the system installation of additional equipment;
- Enacting programs by licensees for early identification of important safety-related issues, and for collecting and analyzing the relevant data, so operating experience can be shared and quickly acted upon;
- Expanding NRC's international activities to share improved knowledge of nuclear safety with other countries within important technical fields (US NRC, 2014).

BHOPAL TRAGEDY

At around midnight on Dec. 3, 1984, a Union Carbide plant manufacturing the pesticide suffered a terrible malfunction, accidentally releasing poisonous gas into the air over Bhopal, India. 45 tons of poisonous methyl isocyanate escaped from the facility (Masi, 2016).

During the late evening hours of 2 December 1984, water was believed to have entered a side pipe and into tank which contained 42 tons of methyl isocyanate (MIC). Introduction of water into the tank began a runaway exothermic reaction, which was accelerated by contaminants, high ambient temperatures and other factors, such as the presence of iron from corroding non-stainless steel pipelines (Eckerman, 2005). By 11:30 p.m., workers in the MIC area were feeling the effects of minor exposure to MIC gas, and began to look for a leak. At 12:40 a.m., the reaction in tank quickly reached a critical state. Temperatures in the tank were indicated off its scale, maxed out beyond 25 °C (77 °F), and the pressure in the tank was indicated at 40 psi. One employee saw a concrete slab above tank E610 crack as the emergency relief valve burst open, and pressure in the tank continued to rise to 55 psi even after atmospheric venting of toxic MIC gas had begun. Direct atmospheric venting should have been prevented or at least partly mitigated by at least three safety devices which were not operating properly, not in use, or inadequately sized (Jones, 2015).

About 30 metric tons of methyl isocyanate entered the atmosphere from the tank in 45 to 60 minutes. This amount increased to 40 metric tons within two hours' time. The gases were blown in a south-eastern direction over Bhopal (Eckerman, 2005).

A UCIL operator triggered the plant's alarm system at 12:50 a.m. as the concentration of gas in and around the plant became problematic to tolerate. Activation of the system triggered two siren alarms: one that signaled inside the UCIL plant, and a second that should announce the dangerous situation to the public and the city of Bhopal. But the public siren briefly sounded at 12:50 a.m. and was quickly turned off, according to company procedure which try to avoid alarming the public around the factory over tiny leaks (Labunska et.al., 1999).

With the lack of appropriate information exchange between pesticides plant and Bhopal authorities, the city's Hamidia Hospital was first told that the gas leak was suspected to be ammonia, then phosgene. They were then told that it was "MIC", which hospital staff had never heard of, had no antidote for, and received no direct information about. Most citizens who were exposed to the MIC gas were first made aware of the leak by exposure to the gas itself, or by opening their doors to investigate the situation, rather than having been instructed to shelter in place, or to evacuate before the arrival of the gas in the first place (Eckerman, 2005).

Shanty towns surrounded the pesticide plant, thus, more than 600,000 people being exposed to the deadly gas cloud that night. The gases stayed low to the ground, causing victims throats and eyes to burn, inducing nausea, and many deaths. Thousands died within hours. More followed over subsequent months – about 15,000 in all. Many of those who survived suffered blindness, organ failure and other awful bodily malfunctions. A shockingly high number of children in the area have been born with all manner of birth defects. In 1989, Union Carbide paid out about half a billion dollars to victims, an amount the afflicted say is not nearly enough to deal with the decades-long consequences (Masi, 2016).

Since then, very little of the toxic residue around the plant has been cleaned up, while the local population suffers from a high rate of birth defects, cancer and other maladies that are tied to the pollution. Many of the families living near the disaster site are poor, urban slum dwellers unable to afford the cost of relocating. In a city where safe drinking water is already a scarcity, they are forced to drink the affected water lying underneath their homes (Masi, 2016).

EXXON VALDEZ OIL SPILL

On March 24, 1989, the American oil tanker the Exxon Valdez collided with the Bligh Reef (Figure 76). This created an oil spill with far reaching consequences in the Prince William Sound in Alaska. Over 11 million gallons of oil were spilled (Disasterium, 2015).

Figure 76: Exxon Valdez Oil Spill.

Source: Disasterium, (2015). Available at: http://www.disasterium.com/10-worst-man-made-disasters-of-all-time/

As US EPA informs, the spill was the largest in U.S. history and tested the abilities of local, national, and industrial organizations to prepare for, and respond to, a disaster of such magnitude. Many factors influenced the cleanup efforts following the spill. The size of the oil spill and its remote location, accessible only by helicopter or boat, made government and industry work difficult and tested existing plans for dealing with such an accident (EPA, 2017).

Since the disaster happened in open navigable waters, the U.S. Coast Guard's On-Scene Coordinator had authority for all activities related to the cleanup operation. His first action was to immediately close the functioning of Valdez Port. The U.S. Coast Guard quickly extended its presence on the scene, and personnel from other Federal agencies also came to help with cleanup. EPA specialists in the use of experimental bioremediation techniques assisted in the spill cleanup. The National Oceanic and Atmospheric Administration was involved in providing weather forecasts for Prince William Sound, making it possible for cleanup team to adapt their methods to changing weather conditions (EPA, 2017). As Environmental Protection Agency admitted (2017), Hubbs Marine Institute in San Diego, California, set up a facility to clean oil from otters, and the International Bird Research Center of Berkeley, California, established a center to clean and rehabilitate oiled waterfowl.

Three methods were tried in the effort to clean up the spill:

- Burning. An experimental burn was conducted during the early stages of the oil spill. The two ships with the boom between them moved gradually throughout the slick until the boom was full of oil. Then the boom was towed away from the slick and the oil was burned. The fire did not endanger the main spill or the Exxon Valdez ship because of the distance separating them (EPA, 2017).
- Mechanical cleanup. Mechanical cleanup was started immediately after the spill. It used booms and skimmers. However, skimmers were not able to use during the first 24 hours following the spill. Thick oil and heavy kelp had a tendency to to clog the equipment. Repairs to damaged skimmers consumed a lot of time. Transporting oil from temporary storage vessels into more permanent tanks was also difficult because of the oil's weight and thickness. Continuous bad weather slowed down the cleanup operations (EPA, 2017).
- Chemical dispersants. An experimental application of dispersants was tried. The dispersants application was considered controversial. Exxon Valdez had less than 4,000 gallons of dispersant available in its terminal, it didn't have any application equipment or aircraft. A private company spread dispersants on March 24, with a helicopter and dispersant bucket. But under existent weather and ocean conditions, there was not enough wave action to mix the dispersant with the oil in the water. Thus, the Coast Guard representatives at the site concluded that the dispersants were not working and so their use was canceled (EPA, 2017).

Hard work to save sensitive areas were begun early in the cleanup. Fragile natural environments were detected, defined according to degree of cleanup, and then ranked for their importance for cleanup. Seal pupping areas and fish hatcheries were given the highest priority, and for these locations special cleaning technologies were approved. Despite the identification of sensitive areas and the rapid start-up of shoreline cleaning, the wildlife rescue was slow. Adequate equipment and tools for this task did not reach the accident area quickly enough (EPA, 2017). Over nearly 500 miles of coastline were polluted. Over a quarter million birds were killed and countless other wildlife. Over 11,000 people aided in the cleanup process (Disasterium, 2015).

As a result of the Exxon Valdez oil disaster, Congress passed the Oil Pollution Act of 1990, which required the Coast Guard to reinforce its regulations on oil tank vessels and oil tank owners and operators. Today, tank hulls provide better spills protection; and communications of vessel captains with vessel traffic centers have improved to make for safer sailing (EPA, 2017).

SHRINKING OF THE ARAL SEA

The Aral Sea is situated in Central Asia, between Kazakhstan and Uzbekistan (Figure 77). Historically, it was the world's fourth largest saline lake, with salinity level of 10 g/l. The two rivers that feed the Aral Sea are the Amu Darya and Syr Darya rivers, respectively reaching the lake from the South and North (Thompson, 2008).

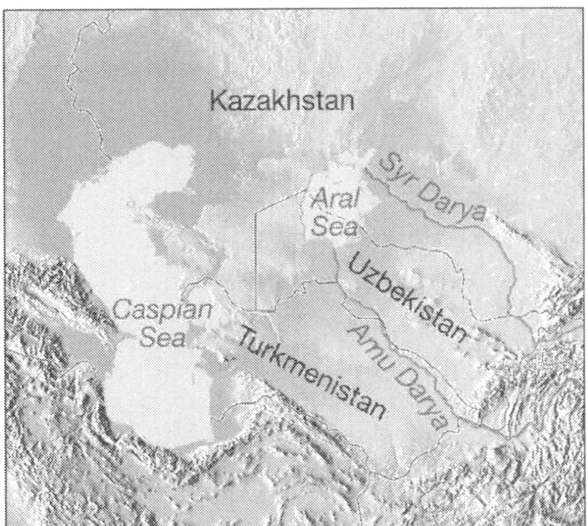

Figure 77: Aral Sea and Amu Darya and Syr Darya Rivers Location.

Source: Thompson, (2008). Available at: http://www.columbia.edu/~tmt2120/introduction.htm

In the 1960s the Soviet government decided to develop agriculture in the desert region surrounding the Sea using irrigation by diverting those rivers back from supplying the Aral Sea basin to supplying desert. By establishing a cotton agricultural program, Soviet government intentionally deprived the Aral Sea of its two main sources of water income, that almost immediately led to less water arriving to the sea. As a result, rivers' water was diverted into canals at the expense of the Aral Sea supply. In addition, the major amount of water was being soaked up by the desert and unashamedly wasted (between 25% and 75% of it, depending on the time period). The dramatical decrease of water level in the Aral Sea started from the 1960s onward (Thompson, 2008).

According to the scientific report (Thompson, 2008), in normal conditions, Aral Sea gets approximately one fifth of its water supply through rainfall, while the rest is delivered to it by the Amu Darya and Syr Darya rivers. Evaporation causes the water level to decline by the same rate that flows into the Sea, making it sustainable as long as inflow is equivalent to evaporation on average. Therefore, the diversion of rivers was the trigger that caused the sea to slowly desiccate over the last 4 decades (Thompson, 2008).

Level of salinity rose from approximately 10 g/l to more than 100g/l in the Southern part of Aral. Salinity of the rivers differs with place and time, as well as according to seasons. Running through the desert, rivers usually collect some salt compounds residues in the ground that result in higher salinity, but may well be lowered again after flowing through irrigated areas. Salinity variability was reduced by constructed dams. Smaller lakes within the Aral Sea were not fed by rivers any more tend to have higher salinity due to evaporation, causing some or all fishes that either survived or had been reintroduced in the 1990s to die. Even re-watering those lakes did not bring the original salinity balance. In 1998, water level was down by 20 m, with a total volume of 210 km^3 compared to 1,060 km^3 in 1960 (Thompson, 2008). Figure 78 shows us the evidence of the Aral Sea disaster during the time.

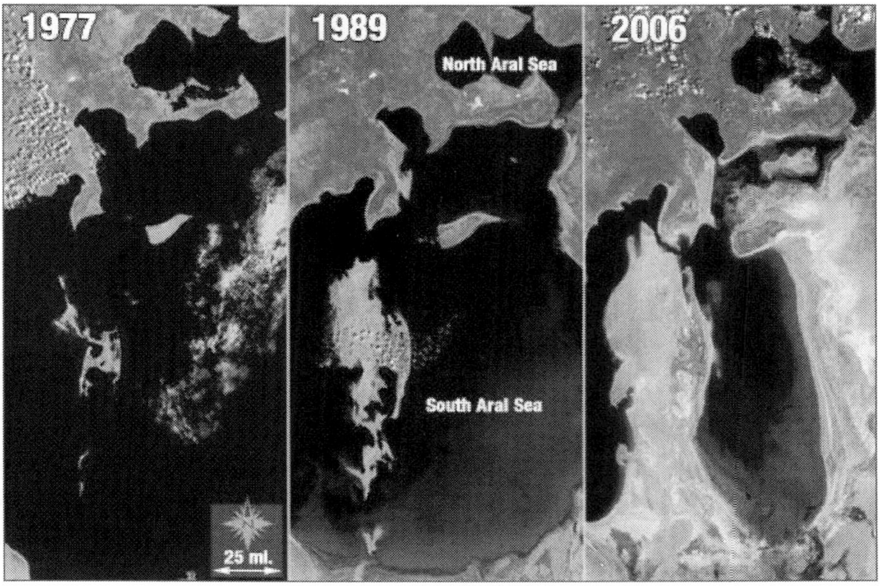

Figure 78: Shrinking of the Aral Sea.

Source: http://www.envis.maharashtra.gov

In early April 2010, United Nations Secretary – General Ban Ki-Moon traveled to Central Asia, where he laid eyes upon a 'graveyard of ships' – rusting fishing trawlers and other vessels stranded in a desert that stretched for miles in all directions (Figure 79). It was the Aral Sea ... or what used to be the Aral Sea. The Aral has shrunk 90%. What was once a vibrant, fish-stocked lake is now a massive desert that produces salt and sandstorms that kill plant life and have negative effects on human and animal health for hundreds of miles around. Scores of large boats sit tilted in the sand – a tableau both sad and surreal (Cruz, 2010).

Figure 79: Graveyard of Ships at the Dried Aral Sea.

Source: Vyacheslav Oseledko/AFP/Getty Images. Available at: http://content.time.com/time/specials/packages/article/0,28804,1986457_1986501_1986451,00.html

Today the 'Aral Desert' is the horrible feedback for human society from our Mother Nature. We must remember at all times that society is responsible for the crisis that has unfolded in and around the Aral, that dramatic changes afflicting the Sea since the 1960s are the result of our environment's reaction to the stresses society has imposed on it (Thompson, 2008).

CHERNOBYL NUCLEAR DISASTER

The April 1986 disaster at the Chernobyl nuclear power plant situated in Ukraine was the result of ambitious but improperly designed experiment on nuclear core cooling process.

On 25 April, before the routine shutdown of reactor # 4, the reactor crew began preparing for a test to determine how long turbines would spin and supply electricity to the main circulating pumps following a loss of main electrical power supply. A series of staff actions, including the disabling of automatic shutdown mechanisms, led the attempted test early on 26 April. By the time when the operator moved to shut down the reactor, the reactor was in an extremely unstable state. As water pumps designed to cool the reactor with circulating water were electrically disconnected, the reactor core became to overheat and then – melt. When water pumps were turned back on, water, which was pumped inside, started to boil.

Operators tried to insert the control rods into the reactor core to slow the nuclear reaction and heat generation. But they were trapped by melted fuel. An idiosyncrasy of the control rods design caused a dramatic power flow as they were inserted into the reactor (World Nuclear Association, 2016). The interaction of very hot fuel with the coolant water resulted in fuel fragmentation along with rapid steam production and an growth of pressure. According to the World Nuclear Association (2016), the design characteristics of the reactor were such that considerable damage to even 3 or 4 fuel assemblies can – and did – result in the destroying of the reactor. The overpressure caused the 1000 t cover plate of the reactor to become partially disconnected, breaking the fuel channels, and blocking all the control rods, which by that time were only halfway down (see Figure 80).

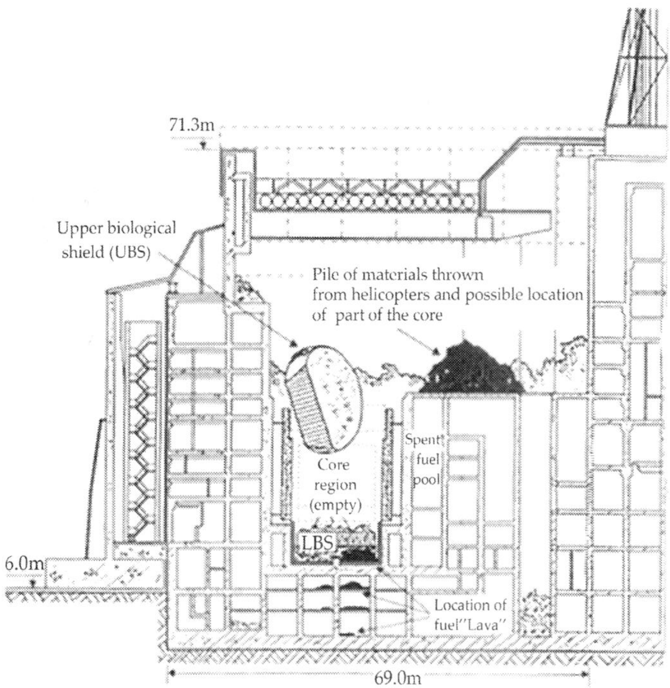

Figure 80: The Damaged Chernobyl Unit 4 Reactor Building.

Source: World Nuclear Association, (2016). Available at: http://www.world-nuclear.org/information-library/safety-and-security/safety-of-plants/chernobyl-accident.aspx

As World Nuclear Association (2016) informs, the intense steam generation spread throughout the whole reactor core (as more water was dumped into the reactor due to the rupture of the emergency cooling circuit) causing a steam explosion and emitting fission products to the atmosphere. About two to three seconds later, the second explosion threw out wreckages of fuel channels and hot graphite. There is no consensus among experts about the origin of this second explosion, but it is likely to have been caused by the production of hydrogen from zirconium-steam reactions (World Nuclear Association, 2016).

1200 tons of graphite and fuel became incandescent and started a number of fires, that caused the main release of radioactivity into the environment (World Nuclear Association, 2016). A total of about 14 EBq (14 x 10^{18} Bq) of radioactivity was released, including iodine-131, caesium-134, cesium-137, strontium-90, xenon-33 and some minor or short-lived nuclides (Loganovsky, 2016).

About 200-300 tons of water per hour was injected into the intact half of the reactor using the supplementary feedwater pumps but this was stopped after half a day owing to the danger of polluted water flowing into and flooding units 1 and 2. From the second to tenth day after the accident, about 5000 tons of boron, dolomite, sand, clay, and lead were dropped on to the burning core by helicopters in an effort to extinguish the blaze and decrease the emission of radioactive particles (World Nuclear Association, 2016).

Chernobyl disaster caused the largest uncontrolled radioactive release into the environment ever recorded for any artificial activity, and enormous amounts of radioactive substances were released into the air for about 10 days causing serious health damage, social and economic disruption for large populations in Belarus, Russia, Ukraine, and other European countries. It is estimated that all the xenon gas, about half of the iodine and cesium, and at least 5% of the remaining radioactive uranium in the Chernobyl 4 reactor core (which had 192 tons of fuel) was released during the accident (World Nuclear Association, 2016). Most of the released radionuclides were deposited close by the place of explosion as dust and debris, but the lighter material was carried by wind over Ukraine, Belarus, Russia and to some extent over Scandinavia and Europe, as it is shown at the Figure 81.

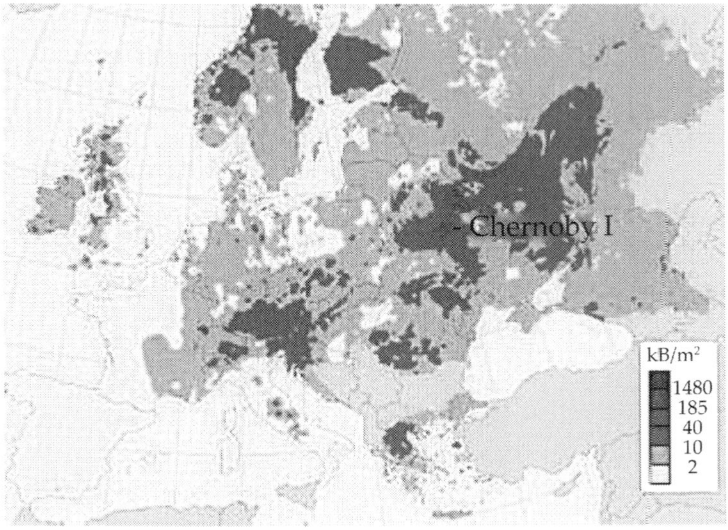

Figure 81: Radioactive fallout from Chernobyl disaster in kBq/m^2.

Source: martinfrost.ws

According to the data of Ukrainian Chernobyl Center, the first day radiation doses people contacted ranged up to 20,000 millisieverts (mSv), resulted in 28 deaths by the end of July 1986. Six of them were firemen who attended the initial fires on the roof of the turbine building (Loganovsky, 2016).

The next task was to reduce the radioactivity cleaning-up at the site of tragedy so that the remaining three reactors could be restarted, and the damaged reactor shielded more permanently. World Nuclear Association (2016) describes that about 200,000 'liquidators' people from all over the Soviet Union were involved in the recovery and clean-up during 1986 and 1987. They received high doses of radiation, averaging close to 100 mSv. Approximately 20,000 of these people received about 250 mSv and a few received 500 mSv. The highest doses were received by about 1000 emergency workers and on-site personnel during the first day of the disaster (World Nuclear Association, 2016).

Initial radiation exposure in contaminated areas was due to short-lived iodine-131; later caesium-137 was the main hazard. About five million people lived in areas of Belarus, Russia and Ukraine contaminated (above 37 kBq/m^2 Cs-137 in soil) and about 400,000 lived in more contaminated areas of strict control by authorities (above 555 kBq/m^2 Cs-137). A total of 29,400 km^2 was contaminated above 180 kBq/m^2 (World Nuclear Association, 2016).

Within the first 48 hours, by April 27, 45,000 residents of Pripyat, plant operators' town, were evacuated. By 14 May, population of 116,000 people within a 30-kilometre radius had been evacuated and later relocated. Most of evacuated residents received radiation doses of close to 50 mSv, although a few received 100 mSv or more. But about 1000 of evacuated people returned unofficially to live within the abandoned polluted zone (Loganovsky, 2016).

In the years following the accident, 220,000 more people were resettled to less contaminated areas, and the initial 30 km radius exclusion zone (2800 km^2) was modified and expanded to cover 4300 km^2. This resettlement was caused by the application of 350 mSv criterion that takes into account projected lifetime radiation dose (World Nuclear Association, 2016).

Several organizations have reported on the impacts of the Chernobyl accident, but all have had problems with evaluation the significance of their observations because of the lack of reliable public health information before 1986 (World Nuclear Association, 2016). Following studies in Ukraine, Russia and Belarus based on national databases evidenced that over one million people was affected by radiation released during Chernobyl accident. By the year of 2000, more than 4000 cases of thyroid cancer had been diagnosed in exposed children. Conifers in about 10 km^2 of forest close to the nuclear plant were killed by the high radiation levels. The exclusion zone has become a unique sanctuary for wildlife where some mutant plants and animals were detected (Loganovsky, 2016).

Nuclear Energy Institute (2015) informs that the crippled Chernobyl 4 reactor now is enclosed in a concrete structure that is growing weaker over time. Officials shut down reactor 2 in 1991 and closed Chernobyl 1 and 3 in 1996 and 2000, respectively. Ukraine and the Group of Eight industrialized nations have agreed on a plan to stabilize the existing structure by constructing an enormous new sarcophagus around it, which is expected to last more than 100 years.

DEEP WATER HORIZON (BP) OIL SPILL

On April 20, 2010, the oil drilling platform Deepwater Horizon, operating in the Macondo Prospect in the Gulf of Mexico, exploded and sank resulting in the death of 11 workers on the Deepwater Horizon and the largest spill of oil in the history of marine oil drilling operations (Pallardy, 2017).

The Deepwater Horizon rig, owned and operated by offshore-oil-drilling company Transocean and leased by oil company British Petroleum (BP), was situated in the Macondo oil prospect in the Mississippi Canyon, a valley in the continental shelf. The oil well was located on the seabed 4,993 feet (1,522 meters) below the surface and extended approximately 18,000 feet (5,486 meters) into the rock. On the night of April 20, a surge of natural gas blasted through a concrete core recently installed by contractor Halliburton in order to seal the well for later use. The core was likely too weak to withstand the pressure because they were composed of a concrete mixture that used nitrogen gas to accelerate curing (Pallardy, 2017).

As Encyclopedia Britannica website informed, once released by the fracture of the core, the natural gas traveled up the Deepwater rig's riser to the platform, where it ignited, killing 11 workers and injuring 17 (Figure 82). The rig overturned and sank on the morning of April 22, rupturing the riser, through which drilling mud had been injected in order to counteract the upward pressure of oil and natural gas. Without any opposing force, oil started to discharge into the ocean. The volume of oil escaping the damaged well—originally estimated by British Petroleum to be about 1,000 barrels per day—was thought by U.S. government officials to have peaked at more than 60,000 barrels per day (Pallardy, 2017).

Figure 82: Fireboats Continue to Battle the Blazing Offshore Oil Rig Deepwater Horizon.

Source: US. Coast Guard—Reuters/Landov, (2010). Available at: https://www.britannica.com/event/Deepwater-Horizon-oil-spill-of-2010

Although BP tried to activate the rig's blowout preventer (BOP), a fail-safe mechanism designed to close the channel through which oil was drawn, the device failed. Efforts to place a containment dome over the largest leak in the broken riser were dissatisfied by the buoyant action of gas hydrates—gas molecules in an ice matrix—formed because of reaction between natural gas and cold water. When an attempt to employ a 'top kill', whereby drilling mud was pumped into the well to stanch the flow of oil, also were unsuccessful, BP in early June decided to use the apparatus called the Lower Marine Riser Package (LMRP) cap. With the damaged riser shorn from the LMRP – the top segment of the blowout preventer – the cap was lowered into place. Though fitted loosely over the BOP and allowing some oil to escape, the cap enabled BP to draw off approximately 15,000 barrels of oil per day to a tanker. The addition of an supplementary collection system including several tools, also tapped into the BOP, increased the collection rate to approximately 25,000 barrels of oil a day (Encyclopedia Britannica, 2017).

Ocean Portal (2016) admits that over the course of 87 days, the damaged Macondo wellhead, located around 5,000 feet beneath the ocean's surface, leaked an estimated 3.19 million barrels (over 130 million gallons) of oil into the Gulf of Mexico—making the spill the largest accidental ocean spill in history (Figure 83).

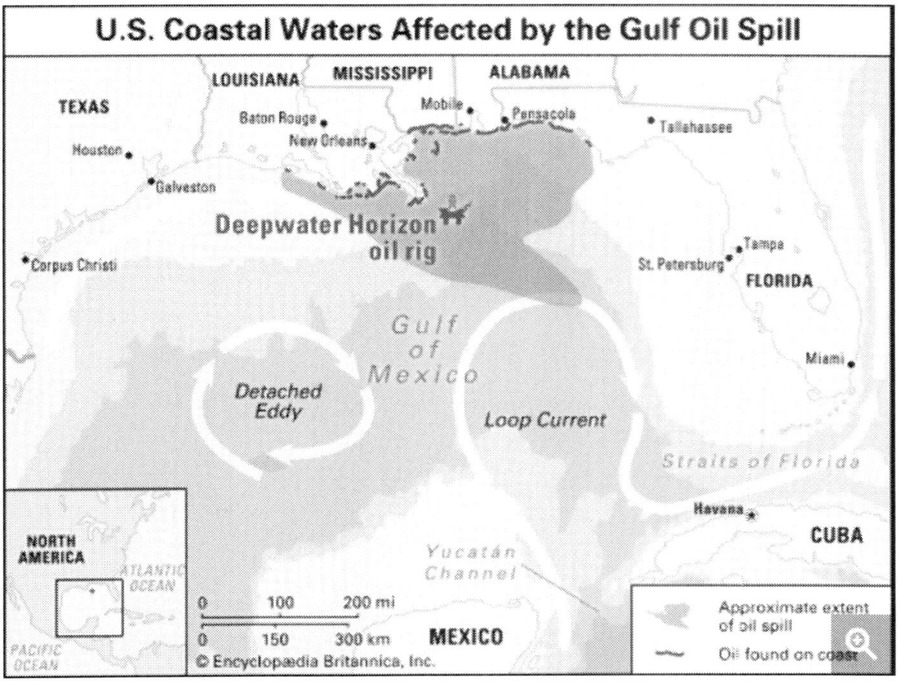

Figure 83: The Effects of the Deepwater Horizon Oil Spill.

Source: Encyclopædia Britannica, Inc. (2016). Available at: https://www.britannica.com/event/Deepwater-Horizon-oil-spill-of-2010

Once the oil came out of the well, it spread throughout the water body. Some of oil floated to the ocean's surface to form oil slicks, which can spread more speedily being pushed by winds. Some hung suspended in the midwater after rising from the wellhead like a chimney and creating several layers of oil, dispersant and seawater mixtures drifting with ocean currents; during the spill a 22-mile long oil plume was detected. This cloud formed influenced chemical dispersants, released into the water to break up the oil so it could be washed away, allowed the oil to mix with seawater and stay suspended below the surface. Finally, some oil sunk to the seafloor by gluing together falling micro material in the water such as bacteria and phytoplankton to form marine snow. As much as 20 % of the spilled oil was estimated to have ended up on top of and in the seafloor, damaging deep sea corals and potentially damaging other natural ecosystems that are unseen at the surface (Pallardy, 2017).

Line of more than 1,000 miles of seashore on the Gulf of Mexico, from Texas to Florida, was damaged by oil spill from the Deepwater Horizon blowout. Much of this territory has been cleaned, but polluted shorelines are taking longer to recover. Conducted research detected acceleration of erosion in these areas. Some immediate impacts to the animals of the Gulf of Mexico could be seen with the naked eye: pelicans and other birds were black with oil, fish belly-up in brown sludge, smothered turtles washed up on beaches. The full understanding of oil spill disaster consequences will take much more time, years of monitoring and research to determine what happened (Ocean Portal, 2016).

Thousands of birds, mammals, and sea turtles were coated with leaked oil. A 2012 study reported that white pelicans that had migrated from the gulf to Minnesota to breed were producing eggs that contained higher levels of compounds that were traceable to the BP spill. Eggs containing traces of pollutants were found in Iowa and Illinois as well (Pallardy, 2017).

Dolphins and sea turtles stranding has increased significantly in the years following the spill. "From 2002 to 2009, the Gulf averaged 63 dolphin deaths a year. That rose to 125 in the seven months after the spill in 2010 and 335 in all of 2011, averaging more than 200 a year since April 2010", reported Reuters in 2015. Typical mechanisms of such widespread fatalities, including morbillivirus and toxins from red tides, were investigated. Researchers to suspected that contaminants from the spill had made cetaceans more vulnerable to other environmental dangers after the unusual incidence of Brucella infection in stranded dolphins was detected (Pallardy, 2017).

Ocean Portal Team (2016) describes that seabirds were initially harmed by crude surface oil which impeded their ability to fly, swim and find food by diving. Some studies showed that Deepwater Horizon accident resulted in hundreds of thousands of seabird losses. Looking beyond the sea, researchers are concentrated on questions how oil may have affected land birds that live in the marshes along the Gulf coast (Ocean Portal, 2016).

Deep-water corals grow very slowly and live for many centuries. Found as deep as 4,000 feet below the surface, corals near the blowout showed signs of tissue damage and were covered by an unknown brown substance, later identified as oil from the spill. Laboratory studies conducted with coral species showed that baby coral exposed to oil and dispersant had lower survival rates and difficulty settling on a hard surface to grow (Ocean Portal, 2016).

The impact of the spill on fish communities is still under the investigation. Lab studies have shown that oil can cause heart defects in the evolving larvae of bluefin tuna and other fish, but we won't know if this occurred in the wild until after those larvae would became adult fish. For years following the spill there were reports of fish with abrasions and abnormalities, and some reports of eyeless and deformed shrimp after the spill (Ocean Portal, 2016).

Economic prospects in the Gulf Coast states were dreadful, as the spill affected many of the industries upon which residents depended. More than a third of federal waters in the gulf were closed to fishing at the peak of the spill, due to fears of pollution. Few tourists were willing to face the prospect of petroleum-sullied beaches, leaving tourism-dependent local population struggling to supplement their incomes (Pallardy, 2017).

FUKUSHIMA NUCLEAR DISASTER

The Great East Japan Earthquake on March 11, 2011 of magnitude 9.0 did considerable damage in the region, and the large tsunami it created caused very much more. As World Nuclear Association (2016) describes, eleven reactors at four nuclear power plants in the region were operating at the time and all shut down automatically when the quake hit. Fukushima Daiichi units 4, 5 and 6 were not operating at the time, but were affected. The main problem initially centered on Fukushima Daiichi units 1-3. Unit 4 became a problem on day five (World Nuclear Association, 2016).

Fukushima Daiichi reactors 1-3 lost power at 3.42 pm, an hour later after the quake, when the entire area was flooded by the 15-metre tsunami. As result, 12 of 13 back-up generators on site were disabled. The same situation happened with heat exchangers for dumping reactor waste heat and decay heat to the sea. The three units lost the aptitude to maintain proper reactor cooling and water circulation process. Electrical switchgear was also disabled. It was detected that no serious damage was caused to the reactors by the earthquake, and the operating units 1-3 were automatically shut down in response to such natural accident, as designed. At the same time, all six external power supply sources were lost due to earthquake damage, so the emergency diesel generators situated in the basements of the turbine area started up. Initially cooling would have been maintained through the main steam circuit bypassing the turbine and going through the condensers (World Nuclear Association, 2016).

The first tsunami wave hit at 3.42 pm, followed by a second wave 8 minutes later. These flooded and destroyed the seawater pumps for both the main condenser circuits and the supporting cooling circuits, notably the Residual Heat Removal cooling system. They also drowned the diesel generators and inundated the electrical switchgear and accumulators, all located in the basements of the turbine buildings. The tsunamis also destroyed and obstructed roads, making outside access problematic.

At 7.03 pm Friday 11 March a Nuclear Emergency was declared, and at 8.50pm the Fukushima Prefecture announced an evacuation order for people within 2 km of the plant. At 9.23 pm the Prime Minister extended evacuation zone to 3 km, and at 5.44 am on 12th he extended it to 10 km. He visited the plant soon after. On Saturday 12th he increased the evacuation zone to 20 km (World Nuclear Association, 2016).

The Fukushima reactors are boiling water reactors (BWR) of an early 1960s design supplied by General Electrics, Toshiba and Hitachi, with a Mark I containment (Figure 84).

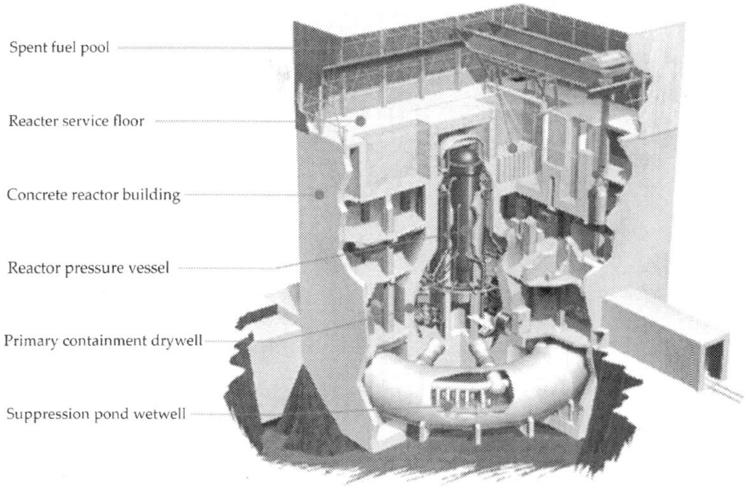

Figure 84: Fukushima Daiichi Boiling Water Reactors Design.

Source: World Nuclear Association, (2016). Fukushima Accident. Available at: http://www.world-nuclear.org/information-library/safety-and-security/safety-of-plants/fukushima-accident.aspx

As World Nuclear Association (2016) reported, when the power failed at 3.42 pm, about one hour after shutdown of the fission reactions, the reactor cores would still be producing about 1.5% of their nominal thermal power, from fission product decay. Without reactor cooling by circulation of water to an outside heat exchanger, this produced a lot of steam in the reactor pressure vessels housing the cores. Produced steam was released into the dry primary containment through safety valves; and later this was accompanied by hydrogen, produced by the interaction of the fuel's very hot zirconium cladding with steam after the water level dropped (World Nuclear Association, 2016).

As pressure started to rise inside the reactor building, the steam was directed into the suppression chamber under the reactor, within the containment, but the internal temperature and pressure nevertheless rose quite rapidly. Water injection started, using the various methods provide for this and finally the Emergency Core Cooling System (ECCS). These systems progressively failed over three days, so from early Saturday water injections to the reactor pressure vessel were conducted with fire pumps, but this required the internal pressures to be relieved initially by venting into the suppression chamber. Seawater injection into unit 1 began at 7pm on Saturday, 12th, into unit 3 – on Sunday,13th and unit 2 – on Monday, 14th. The water level fell to the top of the fuel about three hours after the scram and to the bottom of the fuel 1.5 hours later inside unit 1. The temperature of the exposed fuel into the reactor core rose to some 2800°C so that the central part started to melt after a few hours and by 16 hours after the scram most of it had fallen into the water at the bottom of the reactor pressure vessel. After that temperatures reduced steadily (World Nuclear Association, 2016).

With increasing the pressure, attempts were made to vent the containment. The vented steam, noble gases and aerosols were escorted by hydrogen. Oxidation of the zirconium rods at high temperatures in the presence of steam produces hydrogen exothermically, with this intensifying the fuel decay heat problem. At 3.36 pm on Saturday 12th the hydrogen was mixed with air and ignited that caused the hydrogen explosion on the service floor of the building above unit 1 reactor containment, blowing off the roof and cladding on the top part of the building (World Nuclear Association, 2016). The same happened with units 2 and 3 within the next days. As a result, high volatility fission products including radioactive iodine-131, strontium-90, cesium-134, and cesium-137, were discharged into the environment through the air and deposited in the soil and water (Buesseler et.al., 2012).

One of the primary isotopes that was released by the meltdown of the nuclear reactor was Cs-137. It also was brought to the surface by dry and wet deposition after discharging into the atmosphere. Cs-137 has a half-life of more than 30 years. The soils around Fukushima have been contaminated with depositions of more than 100,000 MBq km^{-2}. The soils in large areas of eastern and northeastern Japan have been strongly contaminated by Cs-137, while western regions of Japan have been sheltered by mountain ranges (Yasunari et.al, 2011). This contamination of the soil has the deleterious effect on stock farming and agriculture, and its potential impacts on human health.

The waters around Fukushima also became the subject to nuclear contamination. According to Buesseler (2011), Cs-137 from the atmosphere was also deposited in the Pacific Ocean. Furthermore, seawater was used to help cool the damaged reactors following the earthquake, and the release of this highly contaminated water led to radioactive discharges directly into the sea. Not only radioactive isotopes were released into the ocean, other nonvolatile activation products and fuel rod materials also were brought into the ocean as they contacted water. A study on the levels of radioactive isotopes in the ocean water near the discharge point has shown significantly elevated concentrations, and the peak of the concentrations in approximately one month after the earthquake, which corresponds with the decision to use seawater to cool the reactors. In terms of radionuclide activities in the surface of the ocean, Fukushima exceeds Chernobyl (Buesseler et.al., 2011). The research of N. Kinoshita and the team (2011) showed that Cs-137 and Cs-134 have been detected throughout waters up to 600 km offshore from Fukushima; cesium isotopes were also detected in zooplankton and mesopelagic fish. The Fukushima disaster is the single largest release of radioactivity into the ocean and one of only two Level 7 nuclear disasters in world history - the other being Chernobyl (Greenpeace, 2016).

According to the Greenpeace, conducted by international research groups studies showed such environmental effects of Fukushima disaster:

- high radiation concentrations in new leaves, and at least in the case of cedar, in pollen;
- increase in growth mutations of fir trees with rising radiation levels;
- heritable mutations in pale blue grass butterfly populations and DNA-damaged worms in highly contaminated areas, as well as apparent reduced fertility in barn swallows;
- decrease in the abundance of 57 bird species with higher radiation levels over a four yeari study;
- high levels of cesium contamination in commercially important freshwater fish; and radiological contamination of one of the most important ecosystems – coastal estuaries (Greenpeace, 2016).

Greenpeace (2016) argues: the "Fukushima nuclear disaster showed us once again that nuclear reactors are fundamentally dangerous. Not only do they cause significant damage to the environment, the health of populations and to national economies, the heavy financial cost of a meltdown is inevitably borne by the public, not by the companies that designed, built, and operated the plants. None of the world's 436 nuclear reactors are immune to human errors, natural disasters, or any of the many other serious incidents that could cause a disaster. Millions of people who live near nuclear reactors are at risk".

APPENDIX A

Guidance for the Assessment of Environmental Incident Severity

Table A-1: Guidance for the Assessment of Environmental Incident Severity

Incident Category	Major	Moderate	Minor
General	Unapproved, enduring, and extensive impact to the environment on or off the Defense Estate (EPBC Act). OR Significant negative impact on management objectives and targets, which requires an emergency commitment of substantial resources (time and/or money) to remediate, OR would take more than 10 years to recover naturally. OR Extensive but short-lived, OR, enduring but localized impact to the environment off the Defense Estate, resulting from an environmental incident that occurred on the Defense Estate.	Unapproved, extensive but short-lived, OR, enduring but localized impact to the environment on or off the Defense Estate (EPBC Act). OR Negative impact on management objectives and targets, which requires a programmed commitment of substantial resources (time and/or money) to remediate, OR would take 2-10 years to recover naturally. OR Short-lived and localized impact to the environment off the Defense Estate, resulting from an environmental incident that occurred on the Defense Estate.	Short-lived and localized, unapproved impact to the environment on or off the Defense Estate (EPBC Act). OR Negative impact on management objectives and targets, which requires a programmed commitment of resources (time and/or money) to remediate, OR would take less than 2 years to recover naturally. OR Potential for impact to the environment off the Defense Estate, resulting from an environmental incident that occurred on the Defense Estate.

Incident resulting from a non-conformance	Moderate or major incident resulting from nonconformance with EPBC Act Referral Conditions (i.e. if a moderate incident occurs and it is also a non-conformance with conditions imposed by an EPBC Act Referral, the incident is upgraded to major severity	Minor incident resulting from non-conformance with EPBC Act Referral Conditions (i.e. if a minor incident occurs and it is also a nonconformance with Conditions imposed by an EPBC Act Referral, the incident is upgraded to moderate severity)	n/a
Incident occurring within approval or conditions (ECC, RSO, SOPs), i.e. impact to the environment above impact expected under the relevant approval or conditions. OR Incident occurring where SOP has not adequately considered environmental impacts	Irreversible and extensive impact to the environment on or off the Defense Estate (EPBC Act) above that approved. OR Significant negative impact on management objectives and targets, above that approved, which requires an emergency commitment of substantial resources (time and/or money) to remediate, OR would take more than 10 years to recover naturally.	Extensive but reversible, OR, irreversible but localized impact to the environment on or off the Defense Estate (EPBC Act) above that approved. OR Negative impact on management objectives and targets, above that approved, which requires a programmed commitment of substantial resources (time and/or money) to remediate, OR would take 2-10 years to ecover naturally.	Reversible and localised damage to the environment on or off the Defence Estate (EPBC Act). OR Negative impact on management objectives and targets, above that approved, which requires a programmed commitment of resources (time and/or money) to remediate, OR would take less than 2 years to recover naturally.

Biodiversity	Loss of or damage to a moderate or large group, or moderate or large area (>100m^2), listed as a Matter of National Environmental Significance under the EPBC Act (i.e. wetlands, threatened species & ecological communities, migratory species, Commonwealth marine areas and Great Barrier Reef Marine Park)	Loss of or damage to a small group or small area (10-100m^2) listed as a Matter of National Environmental Significance under the EPBC Act OR Loss of or damage to a moderate or large group, or moderate or large area (>100m^2), listed under a state listing	Loss of or damage to an individual or very small area (<10m^2) listed as a Matter of National Environmental Significance under the EPBC Act OR Loss of or damage to a small group or small area listed under a state listing OR Loss of or damage to a moderate or large group, or moderate or large area (>100m^2), listed under a local listing
Uncontrolled release – fuels & lubricants	Spill of any size impacts on nationally or internationally significant environmental receptor(s) including soil, air and/or water OR Spill is >50L and impacts on environmental receptor(s) including soil, air and/or water	Spill is 5-50L and impacts on environmental receptor including soil, air and/or water	Spill is <5L and impacts on environmental receptor including soil, air and/or water

Waste	Waste of any size impacts on nationally or internationally significant environmental receptor(s) including soil, air and/or water OR Waste is >50m³ and impacts on environmental receptor(s) including soil, air and/or water	Waste is 5-50m³ and impacts on environmental receptor including soil, air and/or water OR Inappropriate disposal of significant volume of waste (where significant is relative to the type of waste)	Waste is <5m³ and impacts on environmental receptor including soil, air and/or water OR Inappropriate disposal of waste
Heritage	Unapproved, irreversible loss of natural or cultural heritage values from a site with the potential for listing or listed on the World Heritage, National Heritage or Commonwealth Heritage Lists.	Unapproved, extensive but reversible loss of natural or cultural heritage values at a site with the potential for listing or listed on the World Heritage, National Heritage or Commonwealth Heritage Lists. OR Unapproved, irreversible but localized loss of natural or cultural heritage values at a site with the potential for listing or listed on the World Heritage, National Heritage or Commonwealth Heritage Lists.	Unapproved, reversible and localized loss of natural or cultural heritage values at a site with the potential for listing or listed on the World Heritage, National Heritage or Commonwealth Heritage Lists. OR Unapproved, reversible OR localized loss of any natural or cultural heritage values
Resource usage – water, energy, materials	n/a	Excessive, wasteful or inefficient consumption of large quantities of a limited resource including materials, water and energy.	Excessive, wasteful or inefficient consumption of a limited resource including materials, water and energy

Reputation	Damage to Defense's reputation or loss of public confidence at a national or international scale. OR Negative press coverage at a national or international scale. OR Subject of a parliamentary question or ministerial.	Damage to Defense's reputation or loss of public confidence at a local or regional scale. OR Negative press coverage at a local or regional scale	Damage to Defense's reputation or loss of public confidence with a single or small group of organizations or individuals.

Source: Defense National Environmental Standard (DNES), (2014). Available at: http://www.defence.gov.au/estatemanagement/lifecycle/IncidentManagement/Environment/Docs/DNESScheduleAGuidanceforAssessmentofEnvironmentalIncidentSeverity.pdf

APPENDIX B

Synergy between O_3 and Other Pollutants

Table B-1: Synergy between O_3 and Other Pollutants

Subject	**O_3**	**Other exposure**	**Synergistic effect**
Young adult women	480 ppb × 2 hr	270 ppb peroxyacetyl nitrate × 2 hr	Forced expiratory variables
Older men and women	450 ppb × 2 hr	600 ppb NO_2 × 2 hr	Cardiac output and stroke volume

Rat	600 ppb × 4 hr × 2 days	500 μg/m^3 0.3 μm H_2SO_4 × 4 hr × 2 days 500 μg/m^3 0.06 μm H_2SO_4 × 4 hr × 2 days	Alveolar epithelial proliferation Volume density of injured alveoli
Rat	150 ppb × 4 hr × 3 days × 40 weeks	50 μg/m^3 HNO_3 × 4 hr × 3 days × 40 wk	Lung polyamines
Rat	800 ppb × 4 hr	5,000 or 48,000 μg/m^3 resuspended urban PM	Lung cell proliferation
Rat	800 ppb × 4 hr	57,000 μg/m^3 resuspended urban PM	Lung cell proliferation
Guinea pig	1,500 ppb × 1 hr	2 puffs of 33% cigarette smoke	Dynamic lung compliance and resistance
Mouse	500 ppb × 24 hr	30,000 μg/m^3 sidestream cigarette smoke × 6 hr × 3 days	Bronchoalveolar lavage cells and TNFα
Mouse	1,000 ppb × 24 hr	37.5 EU endotoxin × 10 min	Bronchoalveolar lavage IL-1 and IL-6
Rat	500 ppb × 8 hr × 3 days	100 μg intranasal endotoxin × 2 days	Nasal epithelial mucosubstance
Rat	500 ppb × 8 hr × 3 days	100 μg intranasal endotoxin × 2 days	Nasal epithelial mucosubstance
Rat	500 ppb × 8 hr × 3 days	100 μg intranasal endotoxin × 2 days	Nasal epithelial mucosubstance
Rat	500 ppb × 8 hr × 2 days	2 or 20 μg intranasal endotoxin × 2 days	Bronchoalveolar lavage neutrophils
Rat	500 ppb × 8 hr × 3 days	50 μL intranasal 1% ovalbumin × 3 days	Nasal epithelial mucosubstance

Source: Mauderly, J.L., Samet, J.M. (2009). Is There Evidence for Synergy Among Air Pollutants in Causing Health Effects?Environ Health Perspect. 2009 Jan; 117(1): 1–6. Published online 2008 Aug 22. doi: 10.1289/ehp.11654 PMCID: PMC2627851. Available at: https://www.ncbi.nlm.nih.gov/pmc/articles/PMC2627851/

APPENDIX C

Sources of Soil Pollutants and Environmental Effects

Table C-1: Sources of Soil Pollutants and Environmental Effects

Soil pollutant	Potential sources	Poisoning symptoms/effects
Lead (Pb)	Lead paint Mining Foundry activities Vehicle exhaust Construction activities Agriculture activities (pesticide with lead such as lead arsenate that was banned)	Affects the nervous system and memory, growth and development, as well as cognitive development (lower IQs) Learning difficulties Autism in genetically predisposed people Growth reduction and weight loss Impairs the pituitary-thyroid endocrine system Favors osteoporosis
Mercury (Hg)	Mining Incineration of coal Alkali and metal processing Medical waste Volcanoes and geologic deposits (natural sources) Accumulation in plants and vegetables grown on polluted soils	Itching, burning, pain Damage to brain, kidneys, and lungs Pink disease (acrodynia) – skin discoloration (pink cheeks, fingertips, toes), while red cheeks and nose in affected children Desquamation (peeling off in layers of dead skin) High blood pressure and hypersalivation Tachycardia Loss of hair, teeth, nails, photophobia, kidney dysfunction, memory impairment, insomnia in children
Arsenic (As)	Mining Coal-fired power plants Lumber facilities (used as CCA – chromate copper arsenate – in pressure treated wood) Electronics industry Foundry activities Agriculture (Pesticides) Natural accumulation	If ingested – the most specific effects relate to skin pattern changes and cancer (including liver, kidney, bladder, prostate and lung cancer); also at lower doses, the digestive system may be affected with symptoms such as nausea vomiting, stomach irritation, diarrhea, damage to blood vessels If inhaled – skin changes; irritation of throat and lungs, circulatory problems, nervous system disorders
Other metals (Cu, Zn, Ni, etc.)	Mining Foundry activities Construction activities	Depending on the metals, symptoms can include nausea, vomiting, pain, hematemesis (vomiting of blood), hypotension, cramps, diarrhea etc.

PAHs (polyaromatic hydrocarbons)	Coal burning Vehicle emissions Accumulation in plants and vegetables grown on polluted soils Cigarette smoke Wildfires Agricultural burning Wood burning Constructions Industrial plants Accumulation in plants and vegetables grown on polluted soils	Harm to skin, body fluids, and the autoimmune system Eye irritation, nausea, vomiting, diarrhea Cataracts Kidney and liver damage Cancer (skin, lung, bladder, gastrointestinal) Difficulty in reproduction in animals (not yet confirmed in humans)
Herbicides/ Insecticides	Agricultural activities Gardening	Large range of effects, from skin rashes to death

Source: Environmental Pollution Centers. (2017). Available at: https://www.environmentalpollutioncenters.org/soil/

REFERENCES

1. Acid Rains. (2017). US Environmental Protection Agency (EPA). Available at: https://www.epa.gov/acidrain/what-acid-rain
2. Allenby B, Cooper W.E., (1994) Understanding industrial ecology from a biological systems perspective. Total Quality Environ Mgmt. Spring:343–54.
3. Allenby, Brad (2006). The ontologies of industrial ecology (PDF). Progress in Industrial Ecology. Inderscience Enterprises Ltd. 3 (1/2): 28–40. doi:10.1504/PIE.2006.010039.
4. Almasi, A.M., Soque, C., Strandgaard, C.K., Sacchi, R., (2011). Industrial symbiosis Modelling industrial symbiosis to find the potentials and barriers in Aalborg, Denmark. Aalborg University, Energy and Planning department. Available at http://www.symbiosis.dk/sites/default/files/P7%20Industrial%20Symbiosis%20Aalborg%20University%20Christoffer.pdf
5. Ang, C.T., Morad, N., Ismail, N. (2014). Challenges and Possible Drivers of LCA Implementation in Small and Medium Enterprises in Malaysia. Pertanika J. Soc. Sci. & Hum. 22 (2): 335 - 347 (2014) ISSN: 0128-7702 Social Sciences & Humanities. Available at: http://www.pertanika.upm.edu.my/Pertanika%20PAPERS/JSSH%20Vol.%2022%20(2)%20Jun.%202014/01%20Page%20335-348%20(JSSH%200698).pdf
6. Ann Maest 2 , James Kuipers, Kim MacHardy, and Gregory Lawson

Predicted Versus Actual Water Quality at Hardrock Mine Sites: Effect of Inherent Geochemical and Hydrologic Characteristics Poster paper presented at the 7th International Conference on Acid Rock Drainage (ICARD), March 26-30, 2006, St. Louis MO. R.I. Barnhisel (ed.) Published by the American Society of Mining and Reclamation (ASMR), 3134 Montavesta Road, Lexington, KY 40502 pp 1122-1141 DOI: 10.21000/JASMR06021122 . Available at: http://www.asmr.us/Portals/0/Documents/Conference-Proceedings/2006/1122-Maest.pdf

7. Ashton, W., (2008). Understanding the organization of industrial ecosystems, Journal of Industrial Ecology, Vol. 12, No. 1, pp. 34-51.
8. Ashworth W. (1995) The economy of nature. Boston, MA: Houghton Mifflin Company, 89–93.
9. Ayres RU, Ayres L. (1996) Industrial ecology — towards closing the materials cycle. Cheltenham, UK: Edward Elgar, 278–80.
10. Ayres, R., and Kneese, A. (1969). Production, Consumption, and Externalities. American Economic Review, 59 (3), 282–297.
11. Ayres, R.U., (1994). Industrial Metabolism: Theory and Policy. The Greening of Industrial Ecosystems. National Academy Press. 1994. Washington. DC, 23-37. Press. doi: 10.17226/2129. Available at https://www.nap.edu/read/2129/chapter/4#27
12. Bai, X., Peijun Shi, P., and Liu, Y. (2014). Society: Realizing China's Urban Dream. Nature, 509, 158–160.
13. Bannerman, R.T. et al (1993). Sources of Pollutants in Wisconsin Stormwater, Water Science Tech. Vol. 28; No 3-5; pp. 247-259;
14. Bauer, C., Hofer, J., Althaus, H.-J., Del Duce, A., Simons, A. (2015) The environmental performance of current and future passenger vehicles: Life cycle assessment based on a novel scenario analysis framework. Applied Energy, 157, 871–883, doi:10.1016/j.apenergy.2015.01.019.
15. Bauer, C., Treyer, K., (2015). Life Cycle Assessment (LCA). Available at https://www.psi.ch/ta/life-cycle-assessment
16. Begens M., Harper J., Townsend C. (1991). Ecology: Individuals, Populations and Communities. London: Blackwell Press, 28-36.
17. Berkeley Lab. (2008). Life-Cycle Analysis. Available at: http://

newscenter.lbl.gov/2008/04/18/life-cycle-analysis/

18. Betournay, M. C. (2011, April 7). Underground Mining and Its Surface Effects. Retrieved from http://www.fhwa.dot.gov/engineering/geotech/hazards/mine/workshops/iawkshp/betourna2.cfm Bhopal 25 Years Later
19. Bignal, K.L., M.R. Ashmore, & A.D. Headley. 2008. "Effects of air pollution from road transport on growth and physiology of six transplanted bryophyte species". Environmental Pollution. 156: 332-340.
20. Bode, A., Kühner, R., Wisniewski, T., (2011). Identification of economic potentials in production processes: An industrial case study Computers & Chemical Engineering. Volume 35, Issue 12, 14 December 2011, Pages 2591–2601. Available at: https://doi.org/10.1016/j.compchemeng.2011.06.011.
21. Boulding K. (1956). "General Systems Theory—The Skeleton of Science," Management Science 2, no. 3 (April 1956): 197–198.
22. Brattebo, H., Roine, K., Larssæther, S. (1998). State-of-the-art P2005 Industriell økologi. Sintef Teknologiledelse, (In Norwegian).
23. Bright, I., (2008). Air Pollutants, Types and Classification. Tropical-Rainforest-Animals.com. Publication Date: April 2008. Retrieved 05.04.2017 from http://www.tropical-rainforest-animals.com/Air-Pollutants.html
24. Bringezu, S., Schütz, H. (2010). Material use indicators for measuring resource productivity and environmental impacts, Resource Efficiency paper no 6.1 (p. 57). Wuppertal Institute for Climate, Environment and Energy. Berlin.
25. Brown, Richard D.; Ouellette, Robert P.; and Chermisinoff, Paul N. (1983). Pollution Control at Electric Power Stations: Comparisons for U.S. and Europe. Boston: Butterworth-Heinemann.
26. Buesseler, K., Aoyama, M., Fukasawa, M., (2011). Impacts of the Fukushima Nuclear Power Plants on Marine Radioactivity, Environ. Sci. Technol., 45
27. Buesseler, K.O., et al., (2012). Fukushima-Derived Radionuclides in the Ocean and Biota Off Japan, Proc. Nat. Acad. Sci. (USA) 109.
28. Butnariu, A., Avasilcai, S., (2013). Industrial Ecology, A New

Manufacturing Paradigm. Annals of The Oradea University. Fascicle of Management and Technological Engineering. Issue #1, May 2013. Available at: http://imtuoradea.ro/auo.fmte/files-2013-v1/Butnariu%20Anca%201.pdf

29. Cambridge Dictionary. Retrieved 04/05/2017 from http://dictionary.cambridge.org/dictionary/english/industry
30. Canter, L. W (1996). Environmental Impact Assessment. 2nd edn. New York: McGraw Hill.
31. Cardinale, B.J., Duffy, J.E., Gonzalez, A., Hooper, D.V., Perrings, C., Venail, P., Narwani, A., Mace, G.M., Tilman, D., Wardle, D.A., Kinzig, A.P., Daily, G.C., Loreau, M, James B. Grace, J.B., Larigauderie, A., Srivastava, D.S., Naeem, S., (2012). Biodiversity Loss and its Impact on Humanity. Review. Nature | VOL 486 | 7 June 2012. doi:10.1038/nature11148. Available at: http://snre.umich.edu/cardinale/wp-content/uploads/2012/04/Cardinale-et-al-Nature-2012.pdf
32. CENIT de MH (Tecnologías avanzadas para los equipos y procesos de fabricacion de 2015 – Advance technologies for production processes and equipment for the year 2015). (In Spanish).
33. Chandler,D., (2013). Emissionless Method Could Curb a Major Source of Greenhouse Gases. Eco-Friendly Steelmaking. Massachusetts Technological University (MIT), by David June 18, 2013. Available at: https://www.technologyreview.com/s/515411/eco-friendly-steelmaking/
34. Chertow, M. (2000). Industrial Symbiosis: Literature and Taxonomy. Annual Review of Energy and the Environment, 25, 313–337.
35. Chertow, M. R. (2000). Industrial symbiosis: literature and taxonomy, Annual Review of Energy and the Environment, Vol. 25, pp 313-337.
36. Chertow, M. R., and Lombardi D. R. (2005), «Quantifying economic and environmental benefits of colocated firms», Environmental Science and Technology, Vol. 39, No. 17, pp 313-337.
37. Chertow, M.R., (2007). "Uncovering" Industrial Symbiosis. New Haven, 20-21.
38. Climate Change. (2017). US Environmental Protection Agency (EPA). Available at: https://19january2017snapshot.epa.gov/

climatechange_.html

39. Cohen, B. (2005). Journal Ratings and Footprints: A North American Perspective of Organizations and the Natural Environment Journal Quality. Business Strategy and the Environment, 16(1), 64-74
40. Colls, J. (2002). Air Pollution. New York: Spon Press, p. 16-36. Retrieved 04/25/2017 from Questia.com
41. Conserve Energy Future, (2017). What is Green Transportation? Available at: http://www.conserve-energy-future.com/modes-and-benefits-of-green-transportation.php
42. Costa, I. and Ferraro, P. (2010), «A case study of industrial symbiosis using a middle-out approach», Journal of Cleaner Production, Vol. 18, No. 10-11, pp. 984-992.
43. Council of Environmental Quality. (1993). Environmental Quality. Twenty-Third Annual Report. January. Washington D.C.: US Government Printing Office, 151-172.
44. Cruz, G., (2010). And the Earth Cried. Top 10 Environmental Disasters. Time Inc., May 03, 2010. Available at: http://content.time.com/time/specials/packages/completelist/0,29569,1986457,00.html
45. D. Cirtina, N. Ionescu, L.M. Cirtina Environmental Impact Assessement Related to Metallurgical Industry Activities Issn 0543-5846 Journal Metalurgija 55 (2016) 3, 481-484 Available at: http://hrcak.srce.hr/file/226425
46. Dahmus, J.B., Gutowski, T.G. (2007). An environmental analysis of machining. Massachusetts Institute of Technology. 2004. ASME international mechanical engineering congress and RD&D Expo
47. Darmstadter, J., (2010). Decarbonization, Dematerialization, and Environmental Kuznets Curves Available at http://www.rff.org/blog/2010/more-sound-fury-decarbonization-dematerialization-and-environmental-kuznets-curves
48. Darmstadter, J., (2010). More Sound than Fury? Decarbonization, Dematerialization, and Environmental Kuznets Curves, Apr 23, 2010. Available at: http://www.rff.org/blog/2010/more-sound-fury-decarbonization-dematerialization-and-environmental-kuznets-curves
49. Deamer, K., (2016). Scientists determine cause of London›s 1952

'killer fog', Livescience.com December 12, 2016. Available at: http://www.cbsnews.com/news/londons-1952-killer-fog-cause-revealed/

50. Deep Decarbonization Pathways Project (DDPP), (2015). Available at http://deepdecarbonization.org/2015/07/pathways-to-decarbonization-towards-a-2050-low-carbon-australia/
51. Demacon (2009). Pre-Feasibility Assessments on the Establishment of an IDZ. Saldanha Development Zone Pre-Feasibility Analysis – Final Report. October 2009. Available at: http://www.saldanhabay.co.za/pages/IDZ_LED/IDZ/Pre-Feasibility%20Analysis%20Final%20Report_Chapter%208%20&%209.pdf
52. Denesia, M., (2017). SlideSare, Available at: https://www.slideshare.net/MerlynH/industrial-ecology-55184762
53. Design for the Environment Life-Cycle Assessments, (2017). US Environmental Protection Agency (EPA). Available at https://www.epa.gov/saferchoice/design-environment-life-cycle-assessments
54. Disasterium, (2015). 10 Worst Man Made Disasters of All Time. Disasterium. Natural & Man Made Disasters. Available at: http://www.disasterium.com/10-worst-man-made-disasters-of-all-time/
55. Division of Mineral Resources. (1988, January). Solution Salt Mining. In Draft: Generic Environmental Impact Statement on the Oil, Gas, and Solution Mining Regulatory Program (13). Retrieved from http://www.dec.ny.gov/docs/materials_minerals_pdf/dgeisv2ch13.pdf
56. Diwekar, U., (2005). Green process design, industrial ecology, and sustainability: A systems analysis perspective Original Research Article Resources, Conservation and Recycling, Volume 44, Issue 3, June 2005, Pages 215-235 Available at: http://www.sciencedirect.com/science/article/pii/S0921344905000212
57. DocSlides, (2016). Industrial Development. Paralogic-Technologies. Mar 29, 2016. Available at: http://docslide.net/documents/industrial-development-56fae8540d9d7.html
58. Duchina F., Hertwichb E (2003). Industrial Ecology. International Society for Ecological Economics. Online Encyclopaedia of Ecological Economics, May, 2003, 3-11. Retrieved 04/25/2017 from http://isecoeco.org/pdf/duchin.pdf

59. Duflou, J. R., Sutherland, J. W., Dornfeld, D., Herrmann, C., Jeswiet, J., Kara, S., Hauschild, M. and Kellens, K. (2012) «Towards energy and resource efficient manufacturing: A process and system approach», CIRP Annals - Manufacturing Technology, Vol. 61, No. 2, pp. 587-609.

60. Dullinger, S., Essl, F., Rabitsch, W., Erb, K-H., Gingrich, S., Haberl, H., Hülber, K., Jarosik, V., Krausmann, F., Kühn, I., Pysek, P., and Hulme, P. E. (2013). Europe's Other Debt Crisis: The Long Legacy of Human Impact in the Current Extinction Risk of European Taxa. Proceedings of the National Academy of Sciences of the U.S.A., 110, 7342–7347.

61. Eccleston, C. H., (2011). Environmental Impact Assessment: A Guide to Best Professional Practices. Chapter 5. ISBN 978-1439828731

62. Eckerman, I., (2005). The Bhopal Saga—Causes and Consequences of the World›s Largest Industrial Disaster. India: Universities Press. doi:10.13140/2.1.3457.5364. ISBN 81-7371-515-7. Available at: https://docs.google.com/file/d/0B0FqO8XKy9NRZDNzTkZQeVJQbE0/edit?pli=1

63. Ehrenfeld, J. R., (2005). Eco-efficiency Philosophy, Theory, and Tools. Journal of Industrial Ecology, Massachusetts Institute of Technology and Yale University 2005Volume 9, Number 4. Available at: http://www.johnehrenfeld.com/Eco-efficiency%20JIE.pdf

64. Ehrenfeld, J., (1998). Industrial Ecology and Interdisciplinarity: A new Challenge for University Teaching and Research Programs. 3rd Industrial Ecology NTVA-Seminar and Workshop, Trondheim, Norway, 15-16 October 1998.

65. Eisler, R. (1991). Cyanide hazards to fish, wildlife, and invertebrates--A synoptic review. U.S. Fish and Wildlife Service Biological Report 85(1.23), p. 55.

66. Environmental Images. Studyblue. Available at: https://www.studyblue.com/notes/note/n/115-ways-to-go-apes/deck/16284387

67. Environmental Pollution, (2010). Introduction of Environmental Pollution. Sunday, March 28, 2010, Available at: http://environementalpollutionn.blogspot.com/

68. Environmental Protection Agency, (2016). How Mobile Source

Pollution Affects Your Health. Available at: https://www.epa.gov/mobile-source-pollution/how-mobile-source-pollution-affects-your-health

69. Environmental Works. (2008). In Encyclopedia Britannica. Available at: http://library.eb.co.uk/eb/article-214280
70. EPA (1999), Indicators of the Environmental Impacts of Transportation, Center for Transportation and the Environment. Available at: www.itre.ncsu.edu/cte
71. EPA Guidance to Licensees/COA holders on the Notification, Management and Communication of Environmental Incidents. Retrieved 04/25/2017 from https://www.epa.ie/pubs/advice/licensee/Guidance%20to%20licensees.pdf
72. Erb, K.H., Haberl, H., DeFries, R., Ellis, E., Krausmann, F., Verburg, P., Running, S.W., and Smith, W.K. (2012). Pushing the Planetary Boundaries. Science, 338, 1419–1420.
73. Erb, K.H., Kastner, T., Luyssaert, S., Houghton, R.A., Kuemmerle, T., Olofsson, and T., Haberl, H., (2013). Bias in the Attribution of Forest Carbon Sinks. Nature Climate Change, (3), 854–856.
74. European Automobile Manufacture Association (ACEA), (2016). European Automobile Industry Report, 2016/2016. http://www.acea.be/
75. European Communities, 1996. European System of National Accounts (ESA 95)
76. European Environment Agency (EEA), (1998). Ecological and Human impacts. In: Chemicals in the European Environment: Low Doses, High Stakes? Available at: https://www.eea.europa.eu/publications/NYM2/page008.html
77. European System of Accounts (ESA). Retrieved 04/19/2017 from https://stats.oecd.org/glossary/detail.asp?ID=1341
78. Evan, H.Z. (1974) Socio-Economic and labour aspects of pollution control in the chemical industries. International Labour Review. 110 (3), pp. 219-233.
79. Everything Connects, (2013). Soil Pollution. Available at: http://www.everythingconnects.org/soil-pollution.html
80. FAO, (2006). Public Health, Hygiene and Waste Disposal. In:

Fishing Harbour Planning, Construction and Management. UN Food and Agricultural Organization. Available at: http://www.fao.org/docrep/013/i1883e/i1883e12.pdf

81. Farmer, A. (1997). Managing Environmental Pollution. London: Routledge, p. 30-54. Retrieved 05/02/2017 from Questia.com

82. Fedkin, M., (2017). Recycling: open-loop versus closed-loop thinking. EME 807: Technologies for Sustainability Systems. Department of Mechanical and Nuclear Engineering / Dutton e-Education Insitute. College of Earth and Mineral Sciences, The Pennsylvania State University. Available at https://www.e-education.psu.edu/eme807/node/624

83. Finlayson, A., Markewitz, K., and Frayret, J-M. (2014). Postsecondary Education in Industrial Ecology Across the World. Journal of Industrial Ecology, 18(6), 931–941.

84. Fischer-Kowalski, M. (1998). Society's metabolism: The Intellectual History of Materials Flow Analysis, Part I: 1860–1970. Journal of Industrial Ecology, 2(1), 61–78.

85. Fischer-Kowalski, M., and Hüttler, W. (1998). Society's Metabolism. Journal of Industrial Ecology, (2), 107–136.

86. Forman, R., et.al. (2003). Road Ecology: Science and Solutions, Island Press. Available at: www.islandpress.com

87. Friedrich, C., (1998). Environmental Effects. In: Precision Micromanufacturing Processes. Available at: http://www.me.mtu.edu/~microweb/chap2/ch2-8.htm

88. Frosch, R.A., and Gallopoulos, N. E. (1989). Strategies for Manufacturing. Scientific American, 261 (3), 144–152.

89. Fulekar, M.H., (2015). Environmental Impact Assessment, School of Environment and Sustainable Development. Available at http://iced.cag.gov.in/wp-content/uploads/C-30/EIA.pdf

90. Ganguly, (2017). Soil Pollution. In: Types of Pollution. Available at: http://www.tentypesofpollution.com/2016/05/soil-pollution.html

91. Garner A., Keoleian G. A. (1995). Industrial Ecology: An Introduction, 36-37.

92. Geng, Y. (2011). Eco-indicators: Improve China's Sustainability Targets. Nature, 477, 162.

93. Geng, Y., Sarkis, J., Ultiati, S., and Zhang, P. (2013). Measuring China's Circular Economy. Science, 339, 1526–1527.

94. Geron, C. D., Guenther, A. B., and Pierce, T. E. (1994). An improved model for estimating emissions of volatile organic compounds from forests in the eastern United States. Abstract. Journal of Geophysical Research, Vol. 99. Available at: http://www.agu.org/pubs/crossref/1994/94JD00246.shtml

95. Glasmeier, A.K., (2001) Industrial Metabolism. International Encyclopedia of the Social & Behavioral Sciences Elsevier Science Ltd. ISBN: 0-08-043076-7, 7331-7333.

96. Goedkoop, M., Oele, M., Leijting, J., Ponsioen, T., Meijer E. (2016). Introduction to LCA with SimaPro. Language: English Availability: PDF file Available at: https://www.pre-sustainability.com/download/SimaPro8IntroductionToLCA.pdf

97. Gonzalez, A. (2007). Machine Tool Utilisation Phase: Costs and Environmental Impacts with a Life Cycle View. Master of Science Thesis, ISSN 1402-7615 Industrial Ecology, Royal Institute of Technology, Stockholm 2007. Available at: https://www.diva-portal.org/smash/get/diva2:411777/FULLTEXT01.pdf

98. Graedel, T. (1996). On the Concept of Industrial Ecology. Annual Review of Energy and the Environment, 21, 69–98.

99. Greenhouse Effect. (2013). Official Website of The State Delaware. Available at: http://www.dnrec.delaware.gov/climatechange/pages/greenhouse%20effect.aspx

100. Greenpeace, (2016). Fukushima nuclear disaster will impact forests, rivers and estuaries for hundreds of years. Greenpeace report. 4 March 2016. Available at: http://www.greenpeace.org/international/en/press/releases/2016/Fukushima-nuclear-disaster-will-impact-forests-rivers-and-estuaries-for-hundreds-of-years-warns-Greenpeace-report-/

101. Growing Power Bioenergy Solutions from Finland. (2006) Tekes, Finnish Funding Agency for Technology and Innovation. Overview, 10-43

102. Gruber, P. W., Medina, P. A., Keoleian, G. A., Kesler, S. E., Everson, M. P., & Wallington, T. J. (2011). Global lithium availability: A constraint for electric vehicles? Journal of Industrial Ecology,

00(00). Retrieved from http://www.eenews.net/assets/2011/07/27/document_gw_02.pdf

103. Guardian News and Media Limited or its affiliated companies (2017). What›s the carbon footprint of a new car? Available at: https://www.theguardian.com/environment/green-living-blog/2010/sep/23/carbon-footprint-new-car

104. Gudis, C., (2003). Buyways: Automobility, Billboards and the American Cultural Landscape. New York: Routledge.

105. Guidance for Assessment of Environmental Incident Severity. Defence National Environmental Standard Environmental Incident Reporting & Management. Available at: http://www.defence.gov.au/estatemanagement/lifecycle/Incident Management/Environment/Docs/DNESSchedule AGuidancefor Assessment of EnvironmentalIncidentSeverity.pdf

106. Guinée, J.B. (Ed.), Gorrée, M., Heijungs, R., Huppes, G., Kleijn, R., de Koning, A., van Oers, L., Wegener Sleeswijk, A., Suh, S., Udo de Haes, H.A., de Bruijn, J.A., van Duin, R., and Huijbregts, M.A.J. (2002). Handbook on Life Cycle Assessment: Operational Guide to the ISO Standards. Series: Eco-E∞ciency in Industry and Science, 7.

107. Guinée, J.B., Heijungs, R., Huppes, Zamagni, G., Masoni, A.P.R., Buonamici, Ekvall, T., and T., Rydberg (2011). Life Cycle Assessment: Past, Present, and Future. Environmental Science & Technology, 45 (1), 90–96.

108. Habashi, F. (2011). Pollution problems in the metallurgical industry: A review, Journal of Mining & Environment 2(2011)1, 17-26.

109. Harari Y. (2016). Homo Deus: A Brief History of Tomorrow. Harvill Secker, ISBN 9781910701874, 95-96.

110. Harvey, S.K., (2017). Difference Between Industrial Smog and Photochemical Smog. Hearst Seattle Media, LLC. Available at: http://education.seattlepi.com/difference-between-industrial-smog-photochemical-smog-4955.html

111. Hasegawa, A., Tanigawa, K., Ohtsuru, A., Yabe, H., Maeda, M., Shigemura, J., et al. (2015). Health effects of radiation and other health problems in the aftermath of nuclear accidents, with an emphasis on Fukushima, The Lancet, 1 August 2015. Volume 386,

No. 9992, p479–488, Available at: http://www.thelancet.com/pdfs/journals/lancet/PIIS0140-6736(15)61106-0.pdf

112. Health an Environment Alliance. (2011). Transport noise – the health impacts of noise pollution, 30 June 2011. Available at: http://www.env-health.org/news/latest-news/article/transport-noise-the-health-impacts

113. Heijungs, R., Huppes, G.,and Guinée, J.B. (2010). Life Cycle Assessment and Sustainability Analysis of Products, Materials and Technologies. Toward a Scientific Framework for Sustainability Life Cycle Analysis. Polymer Degradation and Stability, 95 (3), 422–428.

114. Henry, J. Glenn, and Heinke, Gary W. (1996). Environmental Science and Engineering. Upper Saddle River, NJ: Prentice-Hall.

115. Herczeg, G., Akkerman, R., Hauschild, M.Z., (2013). Supply chain coordination in industrial symbiosis. Proceedings of the 20th International EurOMA Conference, June 7-12, 2013, Dublin, Ireland, paper SCM-19 1

116. Hertwich, E. (2012). Biodiversity: Remote Responsibility, Nature, 486, 36–37.

117. Hinrichs, Roger A., and Kleinbach, Merlin. (2001). Energy: Its Use and the Environment, 3rd edition. Monterey, CA: Brooks/Cole Publishing Company.

118. History, (2017). Three Mile Island, A&E Television Networks, LLC. Available at: http://www.history.com/topics/three-mile-island

119. Hoekstra, A.Y. and Wiedmann, T.O. (2014.) Humanity's Unsustainable Environmental Footprint. Science, 344 (6188), 1114–1117. Retrieved 04/15/2017 from http://www.sciencemag.org/content/344/6188/1114.abstract

120. Holder, J., (2004), Environmental Assessment: The Regulation of Decision Making, Oxford University Press, New York, 56-79.

121. Hu, M., Kleijn, R., Bozhilova-Kisheva, K., and Di Maio, F.(2013). An Approach to LCSA: The Case of Concrete Recycling. International Journal of Life Cycle Assessment, 1–11.

122. Huppes, G., de Koning, A., Suh, S., Heijungs, R.,van Oers, L.,Nielsen, P., and Guinée, J. B. (2006). Environmental Impacts

of Consumption in the European Union: High-Resolution Input-Output Tables with Detailed Environmental Extensions. Journal of Industrial Ecology, 10(3), 129–146.

123. I^{2}Mine. (2013). Innovative Technologies and Concepts for the Intelligent Deep Mine for the Future. Available at: http://www.i2mine.eu/content/open_access/introduction

124. IIBM Institute of Business Management Industrial Relations (2015). Industrial Ecology. In: Industrial Relations. Available at: https://www.coursehero.com/file/11476866/Industrial-Relations/

125. Indigo Development, (2005). Industrial Ecology. Available at: http://www.indigodev.com/IE.html

126. Industry classification. Investopedia. Retrieved 04/20/2017 from http://www.investopedia.com/terms/i/industry.asp#ixzz4ehNGPR9d

127. Industry Definition. Business dictionary. Retrieved 04/02/2017 from http://www.businessdictionary.com/definition/industry.html

128. International Atomic Energy Agency (2007). IAEA Safety Glossary: Terminology Used in Nuclear Safety and Radiation Protection (PDF). Vienna: IAEA. ISBN 92-0-100707-8. Available at: http://www-pub.iaea.org/MTCD/publications/PDF/Pub1290_web.pdf

129. International Atomic Energy Agency. (2005). Guidebook on environmental impact assessment for in situ leach mining projects. Retrieved from http://www-pub.iaea.org/MTCD/publications/PDF/te_1428_web.pdf.

130. ISO 14040.2 Draft: Life Cycle Assessment - Principles and Guidelines

131. Jedelhauser, M., & Binder, C. R. (2015). Losses and efficiencies of phosphorus on a national level: A comparison of European substance flow analyses. Losses and Efficiencies in Phosphorus Management, 105, Part B, 294–310.

132. Jones, T.R., (2015). Remembering Bhopal: NES Workshop 2015. Available at: http://permalink.lanl.gov/object/tr?what=info:lanl-repo/lareport/LA-UR-15-22804

133. Kalundborg Eco-Industrial Park Symbiosis. Wikipedia. Available at https://en.wikipedia.org/wiki/Kalundborg_Eco-industrial_Park#/

media/File:Kalundborg_Eco-Industrial_Park_Symbiosis_Map.jpgS

134. Kalundborg Symbiosis, 2015. Available at http://www.symbiosis.dk/diagram
135. Kander, A., (2005). Baumol's disease and dematerialization of the economy. Ecological Economics 55: 119–130
136. Kay, J.J., (2002). On Complexity Theory, Exergy and Industrial Ecology: Some implication for Construction Ecology. In Kibert, C., Sendzimir, J., Guy, B. (eds), Construction Ecology: Nature as the Basis for Green Buildings, Spon Press, 72-107.
137. Kennedy, C.A., Ibrahim, N., and Hoornweg, D. (2014). Low Carbon Infrastructure Strategies for Cities, Nature Climate Change, 4, 343–346.
138. Keoleian G.A., Menerey, D., (1993). Life Cycle Design Guidance Manual (Cincinnati: U.S. EPA Risk Reduction Engineering Lab, 14.
139. Keoleian, G.A. (1997). Industrial Ecology of an Automobile - A Life Cycle Perspective. Society of Automotive Engineers (SAE). Warrendale, PA, USA.
140. Kesselmeier, J. et al (2000). Atmospheric volatile organic compounds (VOC) at a remote tropical forest site in central Amazonia. Abstract. Available at: www.Sciencedirect.com
141. Kiełczewski, D. (2008) Konsumpcja a perspektywy zrównoważonego rozwoju, Wydawnictwo Uniwersytetu w Białymstoku, Białystok.
142. Kinoshita, N., et al., (2011). Assessment of Individual Radionuclide Distributions from the Fukushima Nuclear Accident Covering Central-East Japan, Proc. Nat. Acad. Sci. (USA) 108
143. Klein, C., (2012). The Killer Fog That Blanketed London. History in the Headlines, December 6, 2012. Available at: http://www.history.com/news/the-killer-fog-that-blanketed-london-60-years-ago
144. Korhonen J. (2001). Four ecosystem principles for an industrial ecosystem / Journal of Cleaner Production 9 (2001) 253–259, 0959-6526/01/$ - see front matter 2001 Elsevier Science Ltd. All rights reserved. PII: S0959-6526(00)00058-5 Retrieved 04/23/2017 from http://engineering.dartmouth.edu/~d30345d/courses/engs171/industrialecosystem.pdf

145. Kranzberg, G., et.al., (2012). The Environmental Impact of Electricity Production. McMaster University. Available at: http://wbooth.mcmaster.ca/epp/publications/student/Cylma%20Foxton.pdf

146. Krausmann, F., Erb, K-H., Gingrich, S., Haberl, H., Bondeau, A., Gaube, V., Lauk, C., Plutzar, C., and Searchinger, T.D. (2013). Global Human Appropriation of Net Primary Production Doubled in the 20th Century. Proceedings of the National Academy of Sciences of the U.S.A., 110 (25), 10324–10329.

147. Kuipers, J., Maest, A., MacHardy, K., Lawson, G. (2006). Comparison of Predicted and Actual Water Quality at Hardrock Mines: The Reliability of Predictions in Environmental Impact Statements. Available at: www.kuipersassoc.com.

148. Labunska I; Stephenson A; Brigden K; Stringer R; Santillo D; Johnston P.A. (1999). The Bhopal Legacy. Toxic contaminants at the former Union Carbide factory site, Bhopal, India: 15 years after the Bhopal accident (PDF). Greenpeace Research Laboratories, Department of Biological Sciences, University of Exeter, Exeter UK. Available at: http://webdrive.service.emory.edu/users/vdhara/www.BhopalPublications/Environmental%20Health/Greenpeace%20Bhopal%20Report.pdf

149. Lane, B. (2006) Life Cycle Assessment of Vehicle Fuels and Technologies. Camden, United Kingdom.

150. Laner, D., Rechberger, H., & Astrup, T. (2014). Systematic evaluation of uncertainty in material flow analysis. Journal of Industrial Ecology, 18, 859–870.

151. Langford, Terry E. (1990). Ecological Effects of Thermal Discharges. New York: Elsevier Applied Science.

152. LEED Bicyclology. (2017). What is your TRUE Cost to Commute? LEED Bicyclology by LEED Bicycle Solutions, LLC. Available at: https://blog.e-bikerig.com/2015/07/28/what-is-your-true-commute-cost/

153. Lehrke, S. 2004. Memorandum from Stephen Lehrke, Foth & Van Dyke, to Jana Murphy, Flambeau Mining Company. Re: Flambeau Mining Company – 2003 Annual Report Groundwater and Surface Water Trends. January 1, 2004.

154. Lenzen, M., et al. (2012). International Trade Drives Biodiversity Threats in Developing Nations. Nature, 486(7401), 109–112.

155. Leontief, W. (1970). Environmental repercussions and the economic structure: An input–output approach. Review of Economics and Statistics, 52, 262–271.

156. Lifset, R. (1997). A Metaphor, a Field, and a Journal. Journal of Industrial Ecology, vol. 1, no.1, 1997. MIT Press.

157. Lifset, R., and Graedel, T.E. (2002). Industrial Ecology: Goals and Definitions. In Handbook of Industrial Ecology, Ayres, R., and Ayres, L. (Ed). Cheltenham, UK: Edward Elgar.

158. Lifset, R., Graedel, T.E., (1992). Industrial Ecology: Goals and Definitions. Available at http://planet.botany.uwc.ac.za/nisl/ESS/Documents/Industrial_Ecology_Overview.pdf

159. Lowe E. A., Evans L. K. (1995) Industrial ecology and industrial ecosystems. In J. Cleaner Prod., Vol. 3, No. 1-2, pp. 47-53, 1995. Elsevier Science Ltd Printed in Great Britain 0959--6526/95 $10.00 + 0.00

160. Lowe, E. A. (1997). Creating by-product resource exchanges: strategies for eco-industrial parks, Journal of Cleaner Production, Vol. 5, No. 1-2, pp. 57-65.

161. Maciejewski, A. (2007) 'Miliardy w autosalonach', Rzeczpospolita, 17–18.11.2007.

162. Masi, A., (2016). The Legacy of the Bhopal Chemical Leak. Time. Available at: http://content.time.com/time/photogallery/0,29307,1943455_2006658,00.html

163. Massachusetts Institute of Technology. (2015). Environmentally Sensitive 'Green' Mining. Available at: http://web.mit.edu/12.000/www/m2016/finalwebsite/solutions/greenmining.html

164. Mauderly, J.L., Samet, J.M. (2009). Is There Evidence for Synergy Among Air Pollutants in Causing Health Effects?Environ Health Perspect. 2009 Jan; 117(1): 1–6. Published online 2008 Aug 22. doi: 10.1289/ehp.11654 PMCID: PMC2627851. Available at: https://www.ncbi.nlm.nih.gov/pmc/articles/PMC2627851/

165. Mauderly, J.L., Samet, J.M., (2009) Is There Evidence for Synergy Among Air Pollutants in Causing Health Effects? Environmental

Health Perspectives • volume 117 | number 1 | January 2009. Available at: https://ehp.niehs.nih.gov/wp-content/uploads/117/1/ehp.11654.pdf

166. Miemczyk, J., Johnsen, T. E., Macquet, M. (2012), Sustainable purchasing and supply management: a structured literature review of definitions and measures at the dyad, chain and network levels, Supply Chain Management: An International Journal, Vol. 17, No. 5, pp. 478-496

167. Miranda, M., Blanco-Uribe Q., A., HernÃ¡ndez, L., Ochoa G., J., & Yerena, E. (1998). All That Glitters Is Not Gold: Balancing Conservation and Development in Venezuela's Frontier Forests. Retrieved from http://pdf.wri.org/all_that_glitters_is_not_gold.pdf

168. Molycorp (2012). Molycorp innovations. Retrieved from http://www.molycorp.com/technology/molycorp-innovations/

169. Molycorp, Inc. (2010). Molycorp, Inc. (001-34827). Retrieved from http://quote.morningstar.com/stock-filing/Annual-Report/2010/12/31/t.aspx?t=XNYS:MCP&ft=10-K&d=27fc1afdb5a753c7039e5aac54ed08d4

170. Mondal, P., (2016). Environmental Impact Assessment: Definition, Roles and Classification. Available at http://www.yourarticlelibrary.com/environment/environmental-impact-assessment-definition-roles-and-classification/27468/

171. Moreau V., Massard G. (2017). Material and Energy Flow Analysis. Online Publication Date: Mar 2017. DOI:10.1093/acrefore/9780199389414.013.109. Available at http://environmentalscience.oxfordre.com/view/10.1093/acrefore/9780199389414.001.0001/acrefore-9780199389414-e-109

172. Moritz B. (2017). The fourth industrial revolution: it's not all about machines. PwC report. Retrieved 04/21/2017 from http://pwc.blogs.com/ceoinsights/2017/04/the-fourth-industrial-revolution.html

173. Morozan, C., Enache, E., Purice, S., (2002). "Visual Pollution: A New Axiological Dimension Of Marketing?" (PDF). University of Pite, Faculty of Management-Marketing in Economic Affairs Brilla. Available at: http://anale.steconomiceuoradea.ro/volume/2012/n2/125.pdf

174. Muradian, R., Martinez-Alier, J., (2001). Trade and the environment: from a 'Southern' perspective. Ecological Economics 36: 281–297.

175. National Academy of Sciences. (2017). The Automotive Industry. In: Industrial Environmental Performance Metrics Challenges and Opportunities, ISBN 0-309-06242-X, 56-63. Available at: https://www.nap.edu/read/9458/chapter/6#82

176. National Center for O*NET Development (NCOD). 2015. 19-2041.03. O*NET OnLine. Retrieved March 11, 2015, from http://www.onetonline.org/link/summary/19-2041.03.

177. National Industrial Symbiosis Programme (NISP), (2014), Retrieved from http://www.international-synergies.com/projects/nationalindustrial-symbiosis-programme-nisp, accessed October 2014.

178. National Research Council Board on Environmental Studies and Toxicology. (1991). Haze in the Grand Canyon: An Evaluation of the Winter Haze Intensive Tracer Experiment. Washington, D.C.: National Academy Press.

179. NEXT. EU's Sixth framework programme of research, NEXT project (Next generation production systems).

180. Nicolet Minerals Company, 1998. Wisconsin Statute §293.50 Compliance Demonstration. Submitted to Wisconsin Department of Natural Resources (in response to the 1997 Wisconsin Act 171, codified at Wis. Stats. §293.50).

181. Noise and Vibration, (2008). In: Canoga Transportation Corridor Project. Draft Environmental Impact Report (EIR), 6-9. Available at: http://media.metro.net/projects_studies/orange/images/deir/4.9%20Noise%20Section.pdf

182. Norman, D. K., & Raforth, R. L. (1994, December). Cyanide Heap Leaching: A Report to the Legislature. Retrieved from http://www.dnr.wa.gov/Publications/ger_misc_cyanide_heap_leaching_1994.pdf

183. Norton B, Costanza R, Bishop R. C. (1997) The evolution of preferences — why 'sovereign' preferences may not lead to sustainable policies and what to do about it. Ecol Econ 1997; 24:193–211.

184. Norwegian University of Science and Technology (NTNU).

Industrial Ecology Programme (IndEcol). Available at https://www.ntnu.edu/studies/phindok

185. Nuclear Energy Institute, (2014). Lessons From the 1979 Accident at Three Mile Island. Available at: https://www.nei.org/Master-Document-Folder/Backgrounders/Fact-Sheets/The-TMI-2-Accident-Its-Impact-Its-Lessons

186. Nuclear Energy Institute, (2015). Chernobyl Accident and Its Consequences. March 2015. Available at: https://www.nei.org/Master-Document-Folder/Backgrounders/Fact-Sheets/Chernobyl-Accident-And-Its-Consequences

187. O'Rourke, D., Connelly, L., Koshland, C., (1996). Industrial Ecology: A Critical Review International Journal of Environment and Pollution, Vol. 6, Nos. 2/3, pp. 89-112, 1996. Available at: http://web.mit.edu/dorourke/www/PDF/IE.pdf

188. Ocean Portal, (2016). Gulf Oil Spill. Available at: http://ocean.si.edu/gulf-oil-spill

189. OECD (2001). Environmental Outlook for the Chemicals Industry. Organization for Economic Co-Operation and Development. Available at: https://www.oecd.org/env/ehs/2375538.pdf

190. OECD/IEA, (2016). Next-generation wind and solar power. From cost to value. International Energy Agency. Available at: https://www.iea.org/publications/freepublications/publication/NextGenerationWindandSolarPower.pdf

191. Organisation of Economic Co-operation and Development (OECD). (2008). Measuring material flows and resource productivity: Vol. 1. The OECD guide. Paris: OECD.

192. Paliulis, D., (2012). Classification of Relevant Main Pollutants and Emissions Sources as Targets for Plasma Treatment. PlasTEP trainings course and Summer school 2012 Vilnius/Kaunas Part II. VGTU Vilnius 2012. Available at: http://www.plastep.eu/fileadmin/dateien/Events/2012/20716-27_Summer_School/5._Plastep-2012_Paliulis2.pdf

193. Pallardy, R., (2017). Deepwater Horizon oil spill of 2010. Encyclopedia Britannica. Available at: https://www.britannica.com/event/Deepwater-Horizon-oil-spill-of-2010

194. Patrício, J., Kalmykova, Y., Rosado, L., & Lisovskaja, V. (2015).

Uncertainty in material flow analysis indicators at different spatial levels. Journal of Industrial Ecology, 19, 837–852.

195. Paul, J., and Campbell, G. (2011). Investigating rare earth element mine development in epa region 8 and potential environmental impacts (908R11003). U.S. Environmental Protection Agency. Retrieved from website: http://www.epa.gov/region8/mining/ReportOnRareEarthElements.pdf

196. Peterson, L., Hermes, M. (2010). Nuclear Power 2010. Kennesaw State University. Retrieved from http://www.chemcases.com/2003version/nuclear/nc-10.htm

197. Photochemical Smog. (2015). Available at: http://www.markedbyteachers.com/as-and-a-level/geography/photochemical-smog-is-the-most-widely-known-and-perhaps-most-serious-air-pollutant-it-is-formed-in-the-atmosphere-by-the-reaction-between-gaseous-pollutants-nitrogen-oxides-and-hydrocarbons.html

198. Porter M., Van der Lindec, (1996) Green and competitive — ending the stalemate. In: Welford R, Starkey R, editors. Business and the environment. London: Earthscan, 1996.

199. Potential interactions among pollutants. (2009). Environmental Health Perspectives (EHP). 2009 Jan; 117(1): 1–6. Published online 2008 Aug 22. doi: 10.1289/ehp.11654 Available at: https://www.ncbi.nlm.nih.gov/pmc/articles/PMC2627851/table/t2-ehp-117-1/

200. Poulikidou, S., (2010). Identification of the main environmental challenges in a sustainability perspective for the automobile industry. Master of Science Thesis in the Master Programme, Industrial Ecology. Department of Energy and Environment. Division of Physical Resource Theory. Chalmers University of Technology Göteborg, Sweden, 2010, Available at: http://publications.lib.chalmers.se/records/fulltext/136380.pdf

201. Powder Metallurgy Company. (2012). Powder Metallurgy—Intrinsically Sustainable. Available at: http://powdermetallurgyco.com/a-green-technology

202. Pressley, H., (1991). Effects of Transportation on Stormwater Runoff and Receiving Water Quality. Internal Agency Memo, Washington State Department of Ecology. Available at: www.ecy.wa.gov.

203. Principle of Environmental Impact Assessment Best Practice (1999).

International Association for Impact Assessment (IAIA). 1999. PDF, Available at https://web.archive.org/web/20120507084339/http://www.iaia.org/publicdocuments/special-publications/Principles%20of%20IA_web.pdf

204. Radioactive Pollution (2005). In: Environmental Encyclopedia. Radioactive Pollution from Gale. ©2005-2006 Available at: http://www.bookrags.com/research/radioactive-pollution-enve-02/#gsc.tab=0
205. Raghav, (2017). Major Environmental Impact of Transport Development. Available at: http://www.geographynotes.com/articles/5-major-environmental-impact-of-transport-development/249
206. Reck, B.K., and Graedel, T.E. (2012). Challenges in Metal Recycling. Science, 337, 690–695.
207. Reichardt, C. (2008). Heap Leaching and the Water Environment – Does Low Cost Recovery Come at a High Environmental Cost? Retrieved from http://www.imwa.info/docs/imwa_2008/IMWA2008_008_Reichardt.pdf
208. Reijnders, L. 1998. The factor 'X' debate: Setting targets for eco-efficiency. Journal of Industrial Ecology 2(1): 13–22.
209. Responsible Care Global Charter (2016). Emissions of Chemical Industry. Available at: https://www.icca-chem.org/responsible-care/
210. Rising to Global Challenges: 25 Years of Industrial Ecology (2015). International Society for Industrial Ecology (ISIE), Yale School of Forestry & Environmental Studies. Retrieved 04/22/2017 from http://www.is4ie.org/Resources/Documents/ISIE%20Booklet%20 3-16-15.pdf
211. Rodrigue, J.-P., (2017). The Environmental Impacts of Transportation. Available at: https://people.hofstra.edu/geotrans/eng/ch8en/conc8en/ch8c1en.html
212. Royal Commission on Environmental Pollution. Twentieth Report. Transport and the Environment – Developments Since 1994. Presented to Parliament by Command of Her Majesty, September 1997. Available at: http://webarchive.nationalarchives.gov.uk/20110112040753/http:/www.rcep.org.uk/reports/20-transport/documents/1997-20transport.pdf

213. Rueda-Cantuche, J.M., Tukker, A., Evgueni Poliakov, E., Bouwmeester, M., (2009). Towards a global multi-regional environmentally extended input-output database. Ecological Economics 68(7):1928-1937. May 2009 Article (PDF) DOI: 10.1016/j.ecolecon.2008.11.010 Available at: https://www.researchgate.net/publication/46490457_Towards_a_global_multi-regional_environmentally_extended_input-output_database

214. Schaltegger, S. and A. Sturm. 1989. Okologie- ¨ induzierte entscheidungsprobleme des managements. Ansatzpunkte zur ausgestaltung von instrumenten. [Ecology induced management decision support. Starting points for instrument formation.] WWZ-Discussion Paper No. 8914. Basel, Switzerland: WWZ.

215. Schmidheiny, S. 1992. Changing course: A global business perspective on development and the environment. Cambridge, MA: MIT Press.

216. Schuler, D., Buchert , M., Liu , R., Stefanie, G., & Merz, C. (2011, January). Study on rare earths and their recycling. European parliament Retrieved from http://reinhardbuetikofer.eu/wp-content/uploads/2011/01/Rare-earths-study_Oeko-Institut_Jan-2011.pdf

217. Schwieters, N., Moritz, B., (2017). 10 Principles for Leading the Next Industrial Revolution. Business+Technology. Available at: https://www.strategy-business.com/article/10-Principles-for-Leading-the-Next-Industrial-Revolution?gko=f73d3

218. Sehic-Music, N., Goletic, S. Pihura, D., Music, L., Hasanovic, K. (2013). Effects of foundry industry on the environment, Metalurgija 52 (2013) 4, 533-536.

219. SelectUSA, (2016). The Machinery and Equipment Industry in the United States. Overview. Prepared by SelectUSA in collaboration with the International Trade Administration's Industry & Analysis Unit (I&A). Available at: https://www.selectusa.gov/machinery-and-equipment-industry-united-states

220. Shah, R., (2016). Air Pollutants: Classification and Adverse Effects. Available at: http://www.biologydiscussion.com/pollution/air-pollutants/air-pollutants-classification-and-adverse-effects/16725

221. Shayler, H., McBride, M., Harrison, E., (2009). Sources and Impacts of Contaminants in Soils 15 April 2009 Cornell Waste

Management Institute. Available at: http://cwmi.css.cornell.edu/sourcesandimpacts.pdf

222. Shpant (2010). Measuring the water footprint of energy consume. Energy, Technology and Policy. Available at: http://webberenergyblog.wordpress.com/2010/02/21/measuring-the-waterfootprint-of-the-energy-we-consume/
223. Smith, K.R., Ezzati, M., (2005). How Environmental Health Risks Change with Development: The Epidemiologic and Environmental Risk Transitions Revisited. Annual Review of Environment and Resources 30: 291–333
224. Society of Environmental Toxicology and Chemistry (SETAC), Guidelines for Life Cycle Assessment: A "Code of Practice" (Pensacola, Florida: SETAC, 1993).
225. Soil Pollution. (2017). Environmental Pollution Centers. Available at: https://www.environmentalpollutioncenters.org/soil/
226. Southerstone, D., Chappels, H. and van Flied, B. (Eds.) (2004) Sustainable Consumption, Edward Elgar Publishing, Cheltenham.
227. SPECTRA, (2017). 3 Technologies Transforming the Future of Your Commute. February 27, 2017. Available at: https://spectra.mhi.com/3_technologies_transforming_the_future_of_your_commute
228. Stahel, W. R., (2013). A vision of the world in 2030 and a back casting in order to guide actions today. Product-Life Institute. Available at http://www.product-life.org/
229. StatisticaCharts, (2014). Available at: https://www.statista.com/chart/1754/daily-commuting-time-in-us-cities/
230. Stern, D., (2004). The Rise and Fall of the Environmental Kuznets Curve. World Development 32: 1419–1439.
231. Stern, D., (2010). Between estimates of the emissions-income elasticity. Ecological Economics 69: 2173–2182
232. Suter, A.H., (1991). Noise and Its Effects, Administrative Conference of the United States, November 1991, 1-3.
233. The Express Tribune, (2017). Tech world debate on robots and jobs heats up. Technology. By AFP Published: March 26, 2017. Available at: https://tribune.com.pk/story/1365857/tech-world-debate-robots-jobs-heats/

234. The Legacy of the Bhopal Chemical LeakNEXTBACK8 of 11

235. The Rio Declaration On Environment And Development (1992) Retrieved 4/20/2017 from http://www.unesco.org/education/pdf/RIO_E.PDF

236. The World Factbook. Retrieved 04/18/2017 from https://www.cia.gov/library/publications/resources/the-world-factbook/

237. The World in 2050. PwC (2017). Retrieved 04/21/2017 from https://www.pwc.com/gx/en/issues/economy/the-world-in-2050.html

238. TheNational Business (2017). Retrieved 04/18/2017 from http://www.thenational.ae/business/the-life/future-of-industry-automation-is-going-to-disrupt-employment

239. Thermal Pollution. (2017). Pollution Issues. Advameg, Inc. Available at: http://www.pollutionissues.com/Te-Un/Thermal-Pollution.html#ixzz4gPSTKO5E

240. Thompson, (2008). The Aral Sea Crisis. Available at: http://www.columbia.edu/~tmt2120/introduction.htm

241. Tukker, A., Koning, A. de, Wood, R., Hawkins, T., Lutter, S., Acosta, J., Rueda Cantuche, J.M., Bouwmeester, M.C., Oosterhaven, J., Drosdowski, T., and Kuenen, J. (2013). EXIOPOL – Development and Illustrative Analyses of a Detailed Global mr eeSUT/IOT. Economic Systems Research, 25(1), 50–70.

242. U.S. Department of Transportation (1995). Transit Noise and Vibration Impact Assessment. 1995 1st edition; 2006 2nd edition. Federal Transit Administration. Washington D.C. Available at: http://www.fta.dot.gov/documents/FTA_Noise_and_Vibration_Manual.pdf

243. UN Glossary of Statistical Terms. OECD. Retrieved 04/12/2017 from https://stats.oecd.org/glossary/detail.asp?ID=1344

244. UN Joint Group of Experts on the Scientific Aspects of Marine Environmental Protection, Report of the First Session, March 1969, p.5.

245. UNIDO. (2017). Industrial Development Report 2016. The Role of Technology and Innovation in Inclusive and Sustainable Industrial Development. Retrieved 04/22/2017 from https://www.unido.org/fileadmin/user_media_upgrade/Resources/Publications/EBOOK_

IDR2016_FULLREPORT.pdf

246. United Nations, European Commission, International Monetary Fund, Organisation for Economic Co-operation and Development (2003): Handbook of National Accounting: Integrated Environmental Accounting 2003 (SEEA 2003), New York

247. US Environmental Protection Agency (2009). US Office of Solid Waste and Emergency Response. Superfund Soil Screening Guidance: Technical Background Document, EPA/540/R95/128. Accessible at: http://www.epa.gov/oerrpage/superfund/health/conmedia/soil/introtbd.htm

248. US Environmental Protection Agency (2017). Exxon Valdez Spill Profile. Available at: https://www.epa.gov/emergency-response/exxon-valdez-spill-profile

249. US EPA (2017). About the U.S. Electricity System and its Impact on the Environment. Available at: https://www.epa.gov/energy/about-us-electricity-system-and-its-impact-environment

250. US EPA, (1981). The Noise Effects Handbook, Office of Noise Abatement and Control, 4-26.

251. US EPA, (2017). Clean Air Act Title IV - Noise Pollution. Clean Air Act Overview. Available at: https://www.epa.gov/clean-air-act-overview/clean-air-act-title-iv-noise-pollution

252. US EPA. (1995). Pollution prevention guide to using metal removal fluids in Machining operations. Developed for the United States Environmental Protection Agency.

253. US Nuclear Regulatory Commission (US NRC), (2014). Backgrounder on the Three Mile Island Accident, US Nuclear Regulatory Commission December 12, 2014, Available at: https://www.nrc.gov/reading-rm/doc-collections/fact-sheets/3mile-isle.html

254. Van Der Voet H. (2006) Encyclopedia of Environmetrics. Published Online: 15 SEP 2006 DOI: 10.1002/9780470057339.vad015

255. Vernadsky V.I. (1926). Biosphera (The Biosphere), Nauchnoe khimiko-technicheskoye izdatel'stvo (Scientific Chemico-Technical Publishing): Leningrad, 200p.

256. Victoria Transport Policy Institute (2015). Transportation Cost and

Benefit Analysis II – Water Pollution. 10 December 2015, Available at: http://www.vtpi.org/tca/tca0515.pdf

257. Vigon, B. W., Tolle, D. A., Cornary, B. W., Latham, H. C., Harrison, C. L., Bouguski, T. L., Hunt, R. G., and Sellers, J. D., (1993). Life Cycle Assessment: Inventory Guidelines and Principles, EPA/600/R-92/245 (Cincinnati: U.S. Environmental Protection Agency, Risk Reduction Engineering Laboratory, February 1993).
258. Virginia Tech. 2007. Environmental Impact of Electricity Generation and Transmission. Consortium on Energy Restructuring. Available at: http://www.dg.history.vt.edu/ch2/impact.html
259. Wackernagel M, Rees W. (1997) Perceptual and structural barriers to investing in natural capital: economics from an ecological footprint perspective. Ecol Econ 1997;20(1):3–24.
260. Walser T., Limbach L.K., Brogioli R., Erismann E., Flamigni L., Hattendorf B., Juchli M., Krumeich F., Ludwig C., Prikopsky K., Rossier M., Saner D., Sigg A., Hellweg S., Günther D., and Stark W.J. (2012). Persistence of Engineered Nanoparticles in a Municipal Solid Waste Incineration Plant, Nature Nanotechnology, (7), 520–524.
261. Wernet, G., Mutel, C., Hellweg S. & Hungerbühler, K. (2010). The environmental importance of energy use in chemical production. Journal of Industrial Ecology. 15(1): 96-107.
262. What is Soil Pollution? (2017). Conserve Energy Future (CEF). Available at: http://www.conserve-energy-future.com/causes-and-effects-of-soil-pollution.php
263. White, R.M. (1994). Preface In: The Greening of Industrial Ecosystems, B. R. Allenby and D.J. Richards (Ed.). Washington, DC : National Academy Press.
264. Wiedmann, T., Wood, R., Lenzen, M., Minx, J., Guan, D., Barrett, J.,2008. Development of an embedded carbon emissionsindicator — producing a time series of input–output tables andembedded carbon dioxide emissions for the UK by using aMRIO data optimisation system. Report to the UK Departmentfor Environment, Food and Rural Affairs. Defra, London.
265. Wiedmann, T.O., Schandl, H., Lenzen, M., Moran, D.,Suh, S., West, J., and Kanemoto, K. (2013). The Material Footprint of Nations.

Proceedings of the National Academy of Sciences, doi:10.1073/pnas.1220362110.

266. Wikipedia, (2017). Industry. Available at: https://en.wikipedia.org/wiki/Industry#/media/File:Gdp-and-labour-force-by-sector.png

267. Williams, T.G.J.L., Heidrich, O. and Sallis, P.J. (2010) 'A case study of the open-loop recycling of mixed plastic waste for use in a sports-field drainage system', Resources, Conservation and Recycling, 55(2), pp. 118-128.

268. Wisdomjobs. (2012). Pollution: Its Causes, Impact and Prevention. Available at: https://blog.wisdomjobs.com/pollution-its-causes-impact-and-prevention/

269. World Bank, (1998). Types of Environmental Impact. In: Roads and The Environment: A Handbook. Washington, DC: World Bank, Environment Department, 60-65.

270. World Health Organization (WHO), (2017). Ecosystem goods and services for health. Available at: http://www.who.int/globalchange/ecosystems/en/

271. World Health Organization, (2011). Health in the Green Economy. Co-Benefits to Health of Climate Change Mitigation. Transport Sector. Available at: http://www.who.int/hia/hgebrief_transp.pdf

272. World Nuclear Association, (2012). Three Mile Island Accident, March 2001, minor update Jan 2012, Available at: http://www.world-nuclear.org/information-library/safety-and-security/safety-of-plants/three-mile-island-accident.aspx

273. World Nuclear Association, (2016). Chernobyl Accident 1986. World Nuclear Association, (Updated November 2016). Available at: http://www.world-nuclear.org/information-library/safety-and-security/safety-of-plants/chernobyl-accident.aspx

274. World Nuclear Association, (2016). Fukushima Accident. Available at: http://www.world-nuclear.org/information-library/safety-and-security/safety-of-plants/fukushima-accident.aspx

275. World Nuclear Association. (2012 June). In Situ Leach (ISL) Mining of Uranium. World Nuclear Association. Retrieved from http://www.world-nuclear.org/info/inf27.html.

276. Worldatlas. (2016). What Is The Environmental Impact Of The

Mining Industry? Available at: http://www.worldatlas.com/articles/what-is-the-environmental-impact-of-the-mining-industry.html

277. WSDOT (2015) Indirect and Cumulative Impacts. Environmental Manual M 31-11.14 Page 412-1 June 2015

278. Yamano, N., Ahmad, N., 2006. The OECD input–output database. OECD Science, Technology and Industry Working Papers,2006/8. OECD Publishing, Paris, France

279. Yasunari, T.J., et al., (2011). Cesium-137 Deposition and Contamination of Japanese Soils Due to the Fukushima Nuclear Accident,» Proc. Nat. Acad. Sci. (USA) 108

280. Yilmaz, D., (2011). In the Context of Visual Pollution: Effects to Trabzon City Center Silhoutte. The Asian Social Science Journal. 7 (5): 99.

281. Zhang, Y. (2013) Urban metabolism: a review of research methodologies. Environmental pollution,178, pp. 463-473.

282. Zhu, Q., Cote, R.P., (2004), Integrating green supply chain management into an embryonic ecoindustrial development: a case study of the Guitang Group, Journal of Cleaner Production, Vol.12, No. 8-10, pp. 1025-1035.

283. Добровольський В. В. Антропоцентрична модель оцінки сталого розвитку / В. В. Добровольський // Наукові праці : науково-методичний журнал. – Вип. 194. Т. 206. Екологія. – Миколаїв : Вид-во ЧДУ ім. Петра Могили, 2012. – С. 75–78.

284. Клименко, Л.П. Техноекологія / Л. П. Клименко – Одеса: Фонд Екопрінт, Сімферополь: Таврія, 2000. – 542 с.

285. Логановський, К.М., (2016). Медичні наслідки Чорнобильської катастрофи. Науковий центр радіаційної медицини АМН України, Чорнобильський центр з проблем ядерної безпеки, радіоактивних відходів та радіоекології. Available at: http://www.chornobyl.net/ua/index.php?newsid=1181570775

286. Шевченко Л. С. Основи економічної теорії. – Харків: «Право», 2008. – 324 с.

INDEX

B

C

D

E

F

G

I

L